ENERGY POLICIES, POLITICS AND PRICES

DEMAND RESPONSE FOR REDUCED ELECTRICITY CONSUMPTION

A PLAN FOR THE NATION

ENERGY POLICIES, POLITICS AND PRICES

Additional books in this series can be found on Nova's website under the Series tab.

Additional E-books in this series can be found on Nova's website under the E-book tab.

ENERGY POLICIES, POLITICS AND PRICES

DEMAND RESPONSE FOR REDUCED ELECTRICITY CONSUMPTION

A PLAN FOR THE NATION

ADAM R. PETERSON
EDITOR

nova publishers
New York

Library of Congress Cataloging-in-Publication Data

ISBN: 978-1-62948-023-7

Published by Nova Science Publishers, Inc. † New York

Contents

PREFACE

Demand response is a valuable resource for meeting the nation's energy needs. By lowering the peak demand for energy, demand response programs reduce the need to construct new, expensive generation units. However, according to a Federal Energy Regulatory Commission (FERC or Commission) staff report — A National Assessment of Demand Response Potential (National Assessment), submitted to Congress in June 2009 — current demand response programs tap less than a quarter of the total market potential for demand response. The FERC staff has worked with stakeholders to develop a National Action Plan on Demand Response (National Action Plan), which sets out actions to achieve the demand response potential in the United States. Congress required the FERC to develop such a plan in the Energy Independence and Security Act (EISA) of 2007. Because current efforts have missed a significant portion of the cost-effective demand response potential, it is evident that action needs to be taken to either create new programs or expand existing ones where cost-effective. This book provides an overview of the National Action Plan, the process behind its development, and context for understanding it.

Chapter 1 – The Federal Energy Regulatory Commission (FERC or Commission) is required to develop the *National Action Plan on Demand Response* (National Action Plan) as outlined in section 529 of the Energy Independence and Security Act of 2007 (EISA), entitled "Electricity Sector Demand Response." This [National Action Plan is designed to meet three objectives:

1. Identify "requirements for technical assistance to States to allow them to maximize the amount of demand response resources that can be developed and deployed."
2. Design and identify "requirements for implementation of a national communications program that includes broad-based customer education and support."
3. Develop or identify "analytical tools, information, model regulatory provisions, model contracts, and other support materials for use by customers, states, utilities, and demand response providers."

The National Action Plan is intended to be an action plan for the nation for implementation by states, localities, and regions. Although its actions apply to the entire country, the National Action Plan recognizes, as Congress intended, the important role that state and local governments play in developing demand response. The coordinated efforts of federal, state, and local lawmakers and regulators are needed to enable achievement of the nation's demand response potential.

To develop an action plan for the nation, FERC staff solicited and accepted input and participation from a broad range of industry stakeholders; local, state, and federal governing officials including state regulatory utility commissioners; and nongovernmental groups including trade associations. FERC staff conducted extensive outreach to a broad range of entities and used the input to develop *A Discussion Draft on Possible Elements for the National Action Plan* for public comment. A nationally webcast technical conference was held to encourage further input and participation from all interested entities. After the conference, written comments were accepted by the Commission. A draft of the National Action Plan was prepared after reviewing the many ideas that reflect the diverse views of a broad range of entities. This National Action Plan reflects comments received on the *Draft for Comment of the National Action Plan.*

Chapter 2 – The staff of the Federal Energy Regulatory Commission (FERC) and the U.S. Department of Energy (DOE) developed this implementation proposal as required by section 529 of the Energy Independence and Security Act of 2007 (EISA). In particular, this proposal complies with EISA's mandate "to submit to Congress a proposal to implement the [National] Action Plan [on Demand Response], including specific proposed assignments of responsibility, proposed budget amounts, and any agreements secured for participation from State and other participants."

The objective of the proposal is to implement the National Action Plan to achieve the nation's demand response potential. FERC staff and DOE believe that every element of the National Action Plan serves an important part in achieving this potential. This proposal recognizes, however, that it will take public and private actors to effectuate the Plan and, as a result, not every element of that plan may be implemented. As such, this implementation proposal identifies areas where FERC staff and DOE can leverage existing initiatives related to demand response to implement the National Action Plan. At the same time, it recognizes and encourages non-federal government entities to become involved in implementing the National Action Plan through participation in a coalition, or any private or non-federal governmental organizations. FERC staff's primary role with a coalition, or any private or non-federal governmental organizations, will be to monitor, consult with, and provide recommendations regarding the implementation of the National Action Plan. A coalition, or any private or non-federal governmental organizations, should consider the viewpoints of a broad range of demand response stakeholders and private sector organizations. Action under this proposal will not require any new federal appropriations.

In: Demand Response …
Editor: Adam R. Peterson

ISBN: 978-1-62948-023-7

Chapter 1

NATIONAL ACTION PLAN ON DEMAND RESPONSE*

Federal Energy Regulatory Commission Staff

EXECUTIVE SUMMARY

Introduction

The Federal Energy Regulatory Commission (FERC or Commission) is required to develop the ***National Action Plan on Demand Response*** (National Action Plan) as outlined in section 529 of the Energy Independence and Security Act of 2007 (EISA), entitled "Electricity Sector Demand Response."[1] This [National Action Plan is designed to meet three objectives:

1. Identify "requirements for technical assistance to States to allow them to maximize the amount of demand response resources that can be developed and deployed."
2. Design and identify "requirements for implementation of a national communications program that includes broad-based customer education and support."

* This is an edited, reformatted and augmented version of the Federal Energy Regulatory Commission Staff, dated June 17, 2010.

3. Develop or identify "analytical tools, information, model regulatory provisions, model contracts, and other support materials for use by customers, states, utilities, and demand response providers."

The National Action Plan is intended to be an action plan for the nation for implementation by states, localities, and regions. Although its actions apply to the entire country, the National Action Plan recognizes, as Congress intended, the important role that state and local governments play in developing demand response. The coordinated efforts of federal, state, and local lawmakers and regulators are needed to enable achievement of the nation's demand response potential.

To develop an action plan for the nation, FERC staff solicited and accepted input and participation from a broad range of industry stakeholders; local, state, and federal governing officials including state regulatory utility commissioners; and nongovernmental groups including trade associations. FERC staff conducted extensive outreach to a broad range of entities and used the input to develop *A Discussion Draft on Possible Elements for the National Action Plan*[2] for public comment. A nationally webcast technical conference was held to encourage further input and participation from all interested entities. After the conference, written comments were accepted by the Commission.[3] A draft of the National Action Plan was prepared after reviewing the many ideas that reflect the diverse views of a broad range of entities. This National Action Plan reflects comments received on the *Draft for Comment of the National Action Plan*.[4]

Strategic Vision

This National Action Plan calls for the formation of a Coalition to implement the National Action Plan. Coalitions represent a way of harnessing the collective energy of a variety of individual organizations and yielding outcomes that have greater reach and depth than those organizations would achieve acting individually. The Coalition would coordinate the efforts of state/local governing officials, utilities/load-serving entities, demand response providers, regional transmission organizations/independent system operators (RTOs/ISOs), consumer advocates, commercial/industrial customers, the federal government, existing coalitions, and other stakeholders with respect to the strategies and activities described in the National Action Plan.

One or more federal agencies participating in the Coalition (for example, the U.S. Department of Energy (DOE) where appropriate) could direct and (if funds are available) fund implementation of certain identified activities of the National Action Plan. Activities identified in the National Action Plan that lend themselves to being accomplished by private entities would be accomplished using private funding. The work of these private entities would be coordinated with that of federal, state, and local agencies through the Coalition. In addition, it also may be useful for an entity or a federal agency (for example, FERC and DOE) to coordinate and ease implementation by charting the progress of both federal and private efforts to complete the National Action Plan activities, identifying any gaps in implementation and, where appropriate, encouraging and working with other members of the Coalition to fill the gaps.

Strategies and Activities

The National Action Plan consists of strategies and activities to achieve three objectives: technical assistance to states, a national communications program, and the development or identification of tools and materials for use by customers, states, and demand response providers.

1. Technical Assistance to States

The National Action Plan is designed to make available to states the necessary technical assistance to implement effective demand response programs. Such assistance may include:

- Identifying a panel of demand response experts to help inform various constituencies about demand response programs, products, technologies, and incentives and their costs and benefits. Ideally, these experts would be identified as those persons having conducted primary and secondary research in the field (as evidenced by authorship of technical papers), with knowledge of local and regional issues, and the independence to conduct an unbiased assessment.
- Sponsoring or conducting primary and secondary research targeted to various practical aspects of demand response implementation (e.g., sponsorship of technical papers or pilots), and providing technical assistance for demand response.

- Sponsoring and participating in national and regional forums and sessions that will provide information to policymakers, stakeholders, and consumers on efforts to achieve the National Action Plan, as well as research conducted under the National Action Plan.

These steps are meant to be complementary, but the federal agencies, the Coalition, and other private sector actors could implement them as needed, picking and choosing the actions that will best utilize their resources and that have the greatest value for each actor.

2. National Communications Program

EISA requires that the National Action Plan on Demand Response design and identify requirements for implementation of a national communications program that includes broad- based customer education and support. The national communications program should have a multifaceted structure, meaning it would develop and support communications through multiple channels and media to communicate as appropriate with the many audiences for demand response. This plan for a national communications program includes three elements: a Communications Umbrella; local implementation; and direct outreach to states, policymakers, and partners.

The Communications Umbrella, as a national platform, would create and provide a consistent, yet flexible, research-based message framework. This platform would include communications tools, support, and advice that could be adapted for use by interested local demand response implementers.[5] It also would communicate directly with large commercial and industrial customers to inform them of opportunities to participate in demand response programs and the benefits of doing so.

A local implementation strategy could be designed based on this platform to inform residential and commercial customers about the availability and possible benefits of demand response programs, products, technologies, and incentives. Interested local demand response implementers could use the research-based message framework and communications tools and tactics developed by the Communications Umbrella to tailor messages and materials and conduct outreach toward their particular customers.

The national communications program also should have a broad national mission to provide outreach to states, policymakers, and national partners whose support would be needed to develop cost-effective and beneficial demand response, particularly in areas where there are no existing demand response programs or current plans to implement such programs.

These strategies, national and local, would occur concurrently based both on their priority level and on the steps that must occur prior to each stage of the program rollout. Many of the tasks undertaken by the Communications Umbrella, such as foundational market research, would necessarily precede local implementation and direct outreach. However, the experiences of local demand response implementers, national partners, and policymakers will also inform future program development and message framework development under the Communications Umbrella.

3. Tools and Materials

The plan for the development, enhancement, or dissemination of tools and materials involves two general categories of action: 1) a Web-based clearinghouse and 2) tools and methods for assessing the impacts, costs, benefits, and operation of demand response programs.

A Web-based clearinghouse should be established to provide stakeholders with the most current information and analyses on demand response. The Web-based clearinghouse could include existing research as well as new research sponsored by the federal agency or the Coalition under the National Action Plan. The Web-based clearinghouse would serve as a centralized location for collecting all publicly available information on demand response, including regulatory documents, program reports, and other general information. The clearinghouses may be incorporated as part of an existing clearinghouse operated by a federal agency (e.g., Smart Grid Information Clearinghouse) or may be operated and maintained by the Coalition or one of its stakeholders. The clearinghouse should provide information that stakeholders can use to make their own evaluations and judgments about the reports and data that they are able to access through the clearinghouse.

Existing analytical tools and methods will need to be enhanced or developed to aid the expansion of existing demand response programs and the creation of new programs; to advance the use of demand response to support reliable and efficient operations of wholesale transmission, energy, capacity, and ancillary services markets; and to better enable end-use consumers to act optimally while enrolled in a demand response program. The Coalition will provide input and guidance to organizations in enhancing or developing tools and methods. These organizations could engage in new research identifying beneficial characteristics of existing tools or meet through conference venues specifically set up to foster information- sharing on current successful examples.

Introduction

Demand response is a valuable resource for meeting the nation's energy needs. By lowering the peak demand for energy, demand response programs reduce the need to construct new, expensive generation units. However, according to a Federal Energy Regulatory Commission (FERC or Commission) staff report — *A National Assessment of Demand Response Potential* (National Assessment), submitted to Congress in June 2009 — current demand response programs tap less than a quarter of the total market potential for demand response.[6] Because current efforts have missed a significant portion of the cost-effective demand response potential, it is evident that action needs to be taken to either create new programs or expand existing ones where cost-effective.

FERC staff has worked with stakeholders to develop this *National Action Plan on Demand Response* (National Action Plan), which sets out actions to achieve the demand response potential in the United States. Congress required FERC to develop such a plan in the Energy Independence and Security Act (EISA) of 2007.[7] As such, the National Action Plan is intended to be an action plan for the nation, for implementation by states, localities, and regions. Although its actions apply to the entire country, the National Action Plan recognizes, as Congress intended, the important role state and local governments play in developing demand response. The coordinated efforts of federal, state, and local lawmakers and regulators are needed to enable achievement of the nation's demand response potential.

To develop an action plan for the nation, FERC staff and their consultants solicited and accepted input and participation from a broad range of industry stakeholders; local, state, and federal governing officials including state regulatory utility commissioners; and nongovernmental groups including trade associations. FERC staff conducted extensive outreach to a broad range of entities and used the input to develop *A Discussion Draft on Possible Elements for the National Action Plan*[8] for public comment. A nationally webcast technical conference was held to encourage further input and participation from all interested entities. After the conference, written comments were accepted by the Commission.[9] A draft of the National Action Plan was prepared after reviewing the many ideas that reflect the diverse views of a broad range of entities.[10] This National Action Plan reflects comments received on the Draft National Action Plan.

This introduction provides an overview of the National Action Plan, the process behind its development, and context for understanding it.

Legislative Mandate

Congress required FERC to develop the National Action Plan as the second stage of a three- stage process that is outlined in section 529 of EISA, entitled "Electricity Sector Demand Response."[11]

The three-stage process began with the National Assessment, which was submitted to Congress in June 2009. The National Assessment evaluated the national- and state-level potential for demand response that can be achieved within five and 10 years.[12] Additionally, the National Assessment identified barriers that have limited the deployment of demand response. The National Assessment also included specific policy recommendations that, if implemented, could overcome these barriers.

In the second stage, FERC is following Congressional direction to develop the National Action Plan. This document is designed to meet three objectives:

1. Identify "requirements for technical assistance to States to allow them to maximize the amount of demand response resources that can be developed and deployed."
2. Design and identify "requirements for implementation of a national communications program that includes broad-based customer education and support."
3. Develop or identify "analytical tools, information, model regulatory provisions, model contracts, and other support materials for use by customers, States, utilities, and demand response providers."

In the third stage, which will commence once the National Action Plan has been developed, EISA requires "the Commission, together with the Secretary of Energy" to submit a proposal to implement the National Action Plan to Congress. EISA requires the implementation proposal to "include specific proposed assignments of responsibility, proposed budget amounts, and any agreements secured for participation from state and other participants."[13]

Background on Demand Response

Defining Demand Response

Before discussing the actions set out in the National Action Plan, it is necessary to define the term "demand response" as it is used here.

The Commission uses "demand response" to refer to the ability of customers to respond to either a reliability trigger or a price trigger from their utility system operator, load-serving entity, regional transmission organization/independent system operator (RTO/ISO), or other demand response provider by lowering their power consumption. For many years, the term was used to refer to peak clipping actions that were confined to a limited number of hours of the year. As adopted in Order No. 719, the Commission defined "demand response" to mean "a reduction in the consumption of electric energy by customers from their expected consumption in response to an increase in the price of electric energy or to incentive payments designed to induce lower consumption of electric energy."[14]

Demand response can be both dispatchable and non-dispatchable. "Dispatchable demand response" refers to planned changes in consumption that the customer agrees to make in response to direction from someone other than the customer. It includes direct load control of customer appliances such as those for air conditioning and water heating, directed reductions in return for lower rates (called curtailable or interruptible rates), and a variety of wholesale programs offered by RTOs/ISOs that compensate participants who reduce demand when directed for either reliability or economic reasons. This direction to reduce load can be in response to acceptance of the consumer's bid to sell its demand reduction at a price in an organized market (a wholesale price-responsive demand response) or to a retail provider. "Non-dispatchable demand response" refers to programs and products in which the customer decides whether and when to reduce consumption based on a retail rate design that changes over time. This is sometimes called retail price-responsive demand and includes dynamic pricing programs that charge higher prices during high-demand hours and lower prices at other times.

As used in this document, the term "demand response" includes consumer actions that can change any part of the load profile of a utility or region, not just the period of peak usage. As a result of technology innovations and policy directions, new types and applications of demand response are emerging. In particular, consumer response to signals from a utility system operator, load-serving entity, RTO/ISO, or other demand response provider can be deployed to shape any or all parts of a customer's load profile. This concept of demand response encompasses the effect of smart appliances or devices in customer dwellings that can respond automatically in near real-time to the signals of a utility, load-serving entity, or other demand response provider, or to changes in bulk power system conditions such as a change in system frequency. It also includes the smart integration of changeable consumption with variable generation as one method to better enable the addition of new technologies (such as wind farms and rooftop solar systems) to utility systems or customer premises. Demand response also includes deployment of devices that can manage demand as needed to provide grid services such as regulation and reserves, and can also manage power flows into and out of energy storage devices such as plug-in hybrid electric vehicle (PHEV) batteries to provide these same grid services. Demand response can go beyond simple reduction in peak period consumption to include shifting consumption from peak to off-peak hours. For example, the use of thermal energy storage devices may be advanced through the use of time-of-use rates[15] that encourage nighttime charging of home energy storage systems, PHEVs, and all-electric vehicles.

Relationship between Demand Response and Smart Grid

The smart grid and demand response are intrinsically linked in many areas of application. Many of the benefits associated with investment in the smart grid, such as better consumer management of electricity usage in response to prices or signals from grid operators, are, at their core, demand response actions. In turn, investments in some smart grid technologies, such as smart meters, help enable the demand response potential identified in the National Assessment. The Commission acknowledges this linkage in its proposed Smart Grid Policy Statement and Action Plan issued in March 2009:

> Smart Grid advancements will apply digital technologies to the grid, and enable real- time coordination of information from generation supply resources, demand resources, and distributed energy resources (DER). This will bring new efficiencies to the electric system through improved communication and coordination between utilities and with the grid, which will translate into savings in the provision of electric service. Ultimately the smart grid will facilitate consumer transactions and allow consumers to better manage their electric energy costs.[16]

In recognition of this linkage, the actions identified in the National Action Plan are designed to be consistent, and in coordination, with smart grid policies implemented at the federal, state, and local levels. Furthermore, as discussed in Part 2, the National Action Plan calls for developing terminology and messages that emphasize smart energy use, and not artificially separating smart grid actions from demand response actions.

The Geography of Demand Response

To understand the objectives of the National Action Plan, it is useful to review the current state of demand response in the United States and identify the regions where there are opportunities for growth. Such information is contained in the National Assessment, which found that existing demand response programs in the United States have the capacity to offset four percent of current U.S. peak demand.[17] The existing programs are mostly reliability-driven programs — e.g., direct control of end-use loads such as residential air conditioning and curtailable and interruptible rates for large commercial and industrial customers.[18] These programs have been in place for decades.

There is considerable geographical variation in the amount of existing demand response. This variation is driven by several factors, including the rate of load growth, the cost of avoided capacity, and the regulatory disposition at the state level toward demand-side programs. Some areas, such as California, Florida, and New England, have significant demand response activity; others, such as Alaska, Montana, and Wyoming, have little.[19]

If the current level of demand response is preserved through the next decade, the National Assessment estimates that demand response would shave 38,000 megawatts (MW) off U.S. peak demand in the year 2019.[20] If the existing mix of programs were to expand geographically to include regions that currently have little demand response activity and if customer participation rates in every region were to reach the levels representing today's best industry practices, the potential impact of demand response programs would be significantly higher, reaching 82,000 MW, or nine percent of U.S. peak demand.[21]

Indeed, an even more diverse approach to demand response might be pursued in the future. In October 2009, the U.S. Department of Energy (DOE) issued 100 awards totaling $3.4 billion to stimulate the development of the smart grid.[22] The awards, chosen from a list of some 400 proposals, encompass a variety of projects. Some of these projects use demand response based on pricing concepts such as dynamic pricing.

Dynamic pricing has relatively little market penetration today. In 40 of the 50 states (plus the District of Columbia), the National Assessment found that dynamic pricing currently has little to no influence.[23] Either dynamic pricing rates are not being offered in those states or customer participation rates are too low to produce significant impacts. In 10 states, dynamic pricing is estimated to have an impact of one percent or less; in one state, it has an impact of around two percent. However, the National Assessment also found that, with greater inclusion of these price-based demand response programs in the mix of possible options, the total potential for demand response would be far greater. The total potential could range between 14 and 20 percent of peak demand, or 138,000 to 188,000 MW, depending on whether dynamic pricing was deployed on an opt-in basis or opt-out basis.[24]

This snapshot of the state of demand response leads to the conclusion that there is a significant gap between today's levels of demand response and the total potential that resides in the full portfolio of cost-effective options. The clear challenge — and opportunity — is to establish a path forward for realizing this potential consistent with state and federal policies and regulations.

The Objective of the National Action Plan

The National Action Plan aspires to meet the requirements specified in EISA: identify "requirements for technical assistance to States to allow them to maximize the amount of demand response resources that can be developed and deployed;" design and identify "requirements for implementation of a national communications program that includes broad-based customer education and support;" and develop or identify "analytical tools, information, model regulatory provisions, model contracts, and other support materials for use by customers, states, utilities, and demand response providers."[25] To achieve this objective, the National Action Plan calls for two simultaneous, compatible paths of action: 1) enabling development of price-responsive demand response programs, products, technologies, and incentives and 2) facilitating market penetration of newly developing "smart grid" technologies and programs.

With respect to the first path, the National Action Plan seeks to identify technical assistance, communications strategies, and other tools to achieve deployment of the maximum amount of price-responsive demand response identified in the National Assessment. The focus of this path is driven by the results of the National Assessment, which modeled great potential for peak load reductions from price-driven demand response, such as retail dynamic pricing rate designs enabled through advanced metering. The success of these programs will require the approval and full partnership of state and local regulators.

Sidebar 1: Demand Response in Organized Markets

During the past decade, demand response resources have significantly increased their market share in organized markets. FERC and the RTOs/ISOs themselves have acted to eliminate barriers to demand response participation in organized wholesale energy, capacity, and ancillary services markets, and to ensure that demand response resources that are technically capable of providing a service are treated comparably to supply resources offering the service. For example, demand response resources that are capable of providing reserves may participate in the day-ahead and real-time ancillary services markets in the Midwest ISO.[1] Furthermore, demand response resources may offer operating reserves and regulation service in the New York ISO, participate in the California ISO's energy and ancillary services markets, and participate in PJM's Reliability Pricing Model and ISO New England's Forward Capacity Market. Appendix B provides a list of the current opportunities for demand response resources to participate in RTO/ISO markets.

As a means of supporting these initiatives, FERC issued Order No. 719 on October 17, 2008. Through this order, the Commission sought to further address barriers to demand response participation in ISO and RTO markets. Among various reforms, Order No. 719 required system operators to accept bids from qualified demand response resources to provide ancillary services, and eliminated deviation charges during system emergencies to buyers taking less energy in real time than purchased day-ahead.[2] Also, unless prohibited by state or local law, Order No. 719 permitted aggregators of retail customers to bid demand response on behalf of retail customers directly into the organized markets.[3]

Presently, 31,695 MW of demand response are available in ISO and RTO markets, up from 17,146 MW at the end of 2006. Such gains represent 6.6 percent of 2008 peak electricity demand within the regions combined.[4] Demand response capacity resources have nearly tripled in the New England ISO and PJM territories, and resources providing ancillary services accounted for more than 4,000 MW at the end of 2008. Such an infusion of demand response resources has aided in providing greater grid reliability, mitigation of generation market power, and an overall decline in fuel-adjusted power prices in organized wholesale markets.[5]

Moreover, permitting aggregators of retail customers and other curtailment service providers to participate in organized wholesale markets facilitates greater involvement from customers such as universities, big-box retailers, and residential customers.

[1] FERC Staff, *Assessment of Demand Response and Advanced Metering*, at 41 (2008), *available at* http://www.ferc.gov/legal/staff-reports/12-08-demand-response. pdf.
[2] Order No. 719, FERC Stats. & Regs. ¶ 31,281 at P 15.
[3] *Id.*; Order No. 719-A, 128 FERC ¶ 61,059 at P 51. (On rehearing the Commission broadly affirmed its finding, but took a different approach with small utilities.)
[4] ISO/RTO Council, *2009 State of the Markets Report*, at 26 (2009).
[5] *Id.* at 18.

With respect to the second path, the National Action Plan sets forth technical assistance, communications strategies, and other tools to be used to develop and deploy emerging cost-effective smart grid technologies and resources (such as dispatchable and automated demand response) that enable customers to respond to price or other signals. This includes development and deployment of dispatchable demand response as a versatile resource for a variety of applications besides lowering peak demand. Examples include automating demand response, using demand response to provide operating reserves and other ancillary services, and partnering demand response with wind generation and other variable resources when appropriate.

These new and emerging technologies and applications offer benefits that go beyond those that would be derived from the potential estimates in the National Assessment. Further, these technologies can bring the benefits of demand response to all consumers. They also may be helpful by serving as a bridge or facilitating the transition to some form of retail rate design, such as dynamic pricing, for states that choose that path. Therefore, the National Action Plan builds on the successes of demand response initiatives by enabling dispatchable demand response (dispatchable to price and non-price directions) programs, products, technologies, and incentives; fostering the continued deployment of advanced metering initiatives (AMI) and dynamic pricing; employing competitive market forces to develop more customer demand response; and planning for the deployment of many innovative demand response applications that create greater consumer control over energy usage when they create new cost-saving opportunities for consumers. If achieved, these developments should allow business and other customers to make better choices about consumption, which in turn may allow power markets to work better, lower costs, and may benefit the environment.

The National Action Plan is a plan for all parties with a stake in demand response. Stakeholders that have either started to design or already established demand response programs, as well as stakeholders that have an interest in demand response but have not established programs, can benefit from the National Action Plan. The need to appeal to consumers, states, utilities, and a broad range of stakeholders can be met through extensive involvement of these groups throughout the implementation of the National Action Plan.

Process for Developing the National Action Plan

EISA requires the development of the National Action Plan after "soliciting and accepting input and participation from a broad range of industry stakeholders, state regulatory utility commissioners, and non-governmental groups."[26] Also, the Commission is directed to "seek consensus where possible, and decide on optimum solutions to issues that defy consensus."[27] As a result, the process for developing the National Action Plan was designed to gather feedback from interested parties, and the final document reflects feedback from a broad range of industry stakeholders.

Specifically, development of the National Action Plan consisted of the following steps:

1. **Review existing literature.** To initiate the research process and avoid duplication of efforts, FERC staff reviewed the existing literature on demand response, national communications programs, and similar action plans in other areas and industries. The review produced a number of "lessons learned" from these programs. Additionally, this review identified a number of research gaps that have become the focus of some of the specific elements of the National Action Plan.
2. **Conduct stakeholder outreach.** In summer 2008, FERC staff initiated a year-long informal outreach effort involving a diverse group of stakeholders on possible elements of the National Action Plan that would meet the objectives of EISA. This outreach effort consisted of informal interviews with regulators, utilities, consumers, trade organizations, and other stakeholders. The feedback that was received through these efforts contributed to the Discussion Draft.

3. **Develop the Discussion Draft.** This document contained an overarching strategy for the National Action Plan, centered on the creation of a Coalition of stakeholders to steer the effort forward. It also discussed a wide range of possible tactical elements that could be included in the National Action Plan. The Discussion Draft was designed to provide concrete ideas and information to which the various stakeholders could react and provide additional comments. It was made available for public comment when it was posted on the FERC website on October 28, 2009.
4. **Hold a technical conference.** As another means for gathering feedback on the Discussion Draft, FERC staff held a technical conference on November 19–20 in Washington, D.C. The first day of the conference consisted of two panel sessions. A dozen speakers drawn from different facets of the demand response community, including state regulatory utility commissioners and nongovernmental groups, presented their views on the Discussion Draft. The next day, three breakout sessions (moderated by FERC staff) were held on specific elements of the Discussion Draft. The conference closed with a plenary session in which summaries of the breakout sessions were presented. The panel presentations and a summary discussion of the breakout sessions were publicly broadcast over the Internet.[28]
5. **Gather written comments on the Discussion Draft.** In addition to providing feedback at the technical conference, interested parties were encouraged to provide written comments. They filed comments reflecting many points of agreement and some differing opinions.[29] Appendix C lists the organizations that commented during the technical conference or through written comments.
6. **Develop Draft National Action Plan.** FERC staff developed the Draft National Action Plan based on the Discussion Draft and the feedback received through the technical conference and written comments submitted following the technical conference. It was posted on the FERC website on March 11, for public comment.
7. **Review filed written comments on the Draft National Action Plan.** Interested persons provided written comments on the Draft National Action Plan. The filed comments indicated consensus in most areas and disagreement in some others. Appendix C lists the organizations that filed written comments to the FERC on the Draft National Action Plan.

This final National Action Plan is the product of all these consensus-building activities.

Overview of Stakeholder Feedback

Those who filed comments, with few exceptions, were generally supportive of the Draft National Action Plan. In particular, commenters indicate broad support for the development of a Coalition to contribute to the implementation of the National Action Plan. Commenters also agree that cost-effectiveness should be a key consideration in the National Action Plan.

Commenters generally support the message framework development activities under the national communications program, such as developing adaptable, consumer-friendly messages and linking demand response with other smart energy use principles. Submitted comments also reflect general agreement on the importance of conducting foundational market research. Additionally, commenters generally agree that the national communications program should supplement, and not supplant, existing local outreach efforts.

A number of commenters raised issues and expressed different views regarding funding of the National Action Plan. The National Association of State Utility Consumer Advocates (NASUCA) points out that the National Action Plan does not indicate the source of funding for many of the strategies and activities. The American Public Power Association (APPA) states that the federal government should fund the National Action Plan, whereas General Electric Energy (GE) recommends that the Commission identify non-federal funding sources for the initial phase of the National Action Plan. Wal-Mart comments that the Coalition members should not provide funding for the Coalition. The National Action Plan, based on these comments, leaves flexibility for funding from a variety of sources for each strategy and activity.

The comments on both the Draft National Action Plan and on the Discussion Draft indicate a lack of consensus with regard to a demand response grant program. GE and Tendril both provided comments to the Draft National Action Plan strongly supporting the creation of such a program. The Demand Response Coordinating Committee (DRCC) stated that a grant program should be included as one among several options to provide financial support for demand response programs. In the comments on the Discussion Draft, the California Public Utilities Commission (CPUC); Tendril; To the Point; North America Power Producers, Ltd; Energy Curtailment Specialists; APPA; Peak Load Management Alliance (PLMA); Smart Grid Library; and the New York Public Service Commission expressed support for a grant program. In its Discussion Draft comments, the National Rural Electric Cooperative Association (NRECA) stated that a grant program should not be a priority, and Occidental calls the grant proposal overly broad and recommends that grants be restricted to state regulatory commissions or energy agencies to enable them to obtain assistance necessary to facilitate design and implementation of demand response. The Public Utilities Commission of Ohio's (PUCO's) Discussion Draft comments indicate support for a grant program with the caveat that PUCO does not support such a program if funding opportunities are intended to pass through states, because many states are not well structured for grant distribution. Because the majority of comments filed in response to the Discussion Draft and the Draft National Action Plan favor the creation of a demand response grant program, the National Action Plan states that the Coalition should strive to establish such a program, to the extent that the Administration, Congress, and/or private actors decide to make funds available for such a program.

Stakeholders also express divergent views on having two simultaneous, compatible paths of action, as called for in the Draft National Action Plan: 1) enabling price-responsive demand response and 2) facilitating market penetration of newly developing "smart grid" technologies and programs. Both ISO-New England and Wal-Mart support both paths of action. On the other hand, APPA argues that having two paths gives undue weight to smart grid technology, while the Maryland Public Service Commission (Maryland PSC) suggests not explicitly identifying price-responsive demand response as one of the two primary paths, because it is difficult to quantify. Maryland PSC also states that the Draft National Action Plan may be prejudging the value of price-responsive demand response before its cost-effectiveness has been fully evaluated and that the Draft National Action Plan overlooks conflict between wholesale and retail jurisdiction, even though price-responsive demand response hinges largely on deployment of advanced metering initiatives at the retail level. NASUCA comments that the paths of action should maximize the value of cost-effective demand response for consumers. The National Action Plan maintains the two paths of action and emphasizes that both paths take into the account the importance of cost-effectiveness.

Commenters disagree on the role and composition of the Coalition proposed in the draft National Action Plan. APPA supports the draft's approach regarding the Coalition's role to provide input and support to a federal agency implementing the National Action Plan. Xcel states that it is in favor of a Coalition assisting the federal government in implementing the

National Action Plan, but voices its concerns about the effectiveness of a Coalition. In contrast to the approach provided in the draft, the DRCC, the Demand Response and Smart Grid Coalition (DRSG), ISO-New England, National Grid, and San Diego Gas & Electric comment that the Coalition should play the lead role in implementation of the National Action Plan.

Though commenters disagree on the role of the Coalition, DRCC, DRSG, and Ice Energy each support forming a coalition of coalitions. DRCC states that it has already begun to form a coalition of coalitions, and National Grid states that the DRCC already provides a national forum for stakeholders and would be an appropriate candidate to serve as a Coalition leader. Edison Electric Institute (EEI) and Southern Company state that the National Action Plan should implement a decentralized approach whereby several regional coalitions are created to develop and oversee execution of a demand response strategy.

In light of these comments expressing various thoughtful views, the National Action Plan has been revised so that Coalition members, collectively or individually, would implement various National Action Plan activities. It provides that responsibility for particular activities would be determined and coordinated through Coalition discussions. The Coalition would provide an opportunity for stakeholders to share insights and experiences and to identify common goals, areas of mutual concern or need, and current gaps.

Commenters voice differing opinions on the Coalition's composition. New York Transmission Owners (NYTO) argues that the Coalition should be made up of a broad group of organizations representing a variety of interests and regions. NASUCA emphasizes that the Commission should ensure that the consumer voice is represented in the Coalition and states that the voluntary nature of the Coalition and the number of individuals involved could lead to organizational problems. Xcel states that Coalition membership should be limited so that the Coalition will be an effective liaison between the federal agency and stakeholders, and states that its membership should include those with qualitative/technical expertise and those with policy/regulatory expertise. Because of the need to involve all stakeholders, the National Action Plan provides that the Coalition will be composed of volunteers who represent federal, state, and local governments; utilities; load-serving entities; and other demand response providers, RTOs and ISOs, commercial and industrial customers, consumer advocacy organizations, demand response coalitions, and other interest groups.

There is no consensus among commenters regarding the role of the federal government. Several commenters state that the federal agency should oversee, support, advise, and facilitate Coalition implementation of the National Action Plan. San Diego Gas & Electric (SDG&E), on the other hand, voices concern that the draft envisions too large a role for a federal agency. Despite such concerns, commenters state that DOE is the logical and appropriate federal agency to be involved in the National Action Plan's implementation. After consideration of the comments, the National Action Plan now states that one or more federal agencies will be part of the Coalition, and have a unique role, but not necessarily a lead role, in implementation of the National Action Plan.

At least one commenter disagrees with some of the Draft National Action Plan's strategies and activities related to technical assistance to states. The CPUC suggests that there might be no need for the Coalition to provide technical advice to states, because private organizations can do so through state and local proceedings. The Illinois Commerce Commission, along with several others, believes technical assistance to states is critical, however. The National Action Plan retains these actions for technical assistance to states in part because it is a clear statutory requirement of EISA, as well as because of the strong support for efforts regarding technical assistance to states at the October technical conference and in written comments on the Discussion Draft.

Several commenters take opposing views on the need for a national educational campaign with a broad message on demand response as part of the national communications program. In filed comments responding to the Discussion Draft, Southern California Edison Company, Occidental, SDG&E, and Florida Power & Light voiced their opposition to the creation of a national demand response message communicating directly to the mass market. NASUCA and the NRECA expressed the view that a uniform national message would provide incorrect information to many consumers and could do more harm than good. NRECA, CPUC, and others also stated that a national message could override or confuse existing local communications programs that have succeeded in developing and promoting locally focused and locally appropriate demand response programs for consumers. Based on this input, the Draft National Action Plan proposed developing a research-based message framework, tools, and support materials that local demand response implementers can tailor to supplement or enhance their current communications efforts rather than providing a national demand response message. Wal-Mart, NRECA, and APPA support the message framework approach taken in the national communications program section as envisioned in the Draft National Action Plan. EnerNOC's comments to the Draft National Action Plan point out that a blanket advertising campaign at the national level would not be realistic given the variation among states, regions, and customers.

However, in its response to the Draft National Action Plan, DRCC comments that the final plan should include a comprehensive communications component that attempts to undertake high-level awareness and educational communications in addition to efforts that support local communications. This recommendation conflicts with comments received in response to the Discussion Draft. EEI also supports broad-based national public education messages, but states that national messages should not supplant the communications efforts from local or regional level program providers to end-use customers. Because of the broad, but non- unanimous opposition to national messages, the National Action Plan retains the research- based message framework approach whereby the national communications program would provide to local demand response implementers[30] communications tools, support, and advice that local demand response implementers can then adapt to their needs.

During the November 2009 staff technical conference, several stakeholders emphasized the importance of outreach to commercial and industrial customers. Hess supported broad communications to commercial and industrial customers stating that these customers will deliver the most impact early in the National Action Plan. Alcoa stated that large industrial customers can potentially deliver large amounts of demand response. PLMA also stated that the commercial and industrial sector is capable of substantial demand response contributions and suggested facilitating their participation through educational support and coordinating outreach with existing delivery channels. However, several commenters on the Draft National Action Plan, such as American Electric Power, the CPUC, EEI, Southern Company, and SDG&E, argue against direct communications with large commercial and industrial customers as part of the national communications program. The National Action Plan retains the outreach and support materials effort to large commercial and industrial customers that is needed to develop the large demand response potential of this unique customer class. However, to reflect commenters' concerns, the National Action Plan calls for the Coalition to coordinate outreach to these customers, removes the role of a federal agency in overseeing outreach to these customers, and focuses the outreach on providing general information and support materials.

The comments reflect some difference of views regarding the development of tools for end- use customers. DRCC believes that it is unlikely that tools will need to be developed for end- use customers, because private sector innovation is currently addressing this need. National Grid states that there is a need for additional tools for customers to use in analyzing and implementing demand response programs. Because of the EISA requirement to identify and develop such tools and materials as well as the large untapped national potential for demand response by end-use customers and the widespread belief by many stakeholders that better tools are needed to facilitate their participation in demand response programs, the National Action Plan provides for the development of such tools.

Structure of the National Action Plan

The rest of the National Action Plan is divided into two main parts.

Part 1 provides an overview of the strategic vision of the National Action Plan. It begins with a description of the Coalition-based approach as the framework for implementing the elements of the National Action Plan. It then describes basic principles of the Coalition. Finally, Part 1 presents case studies of how the Coalition approach has succeeded in other applications.

Part 2 sets out the elements of the National Action Plan. These actions are organized into three categories: technical assistance to states; the national communications program; and development or identification of tools and materials for use by customers, states, and demand response providers.

PART 1: STRATEGIC VISION OF THE NATIONAL ACTION PLAN

1.1. The Coalition

This National Action Plan calls for the formation of a Coalition to coordinate the implementation of the National Action Plan. Coalitions represent a way to harness the collective energy of a variety of individual organizations, yielding outcomes that have greater reach and depth than those organizations would achieve acting individually. The Coalition would coordinate and combine the efforts of state/local governing officials, utilities/ load-serving entities, RTOs/ISOs, consumer advocates, commercial/industrial customers, the federal government, existing coalitions, and other stakeholders with respect to the strategies and activities described in the National Action Plan.

Why Should a Coalition Be Involved in Implementing the National Action Plan?

The process of generating electricity and delivering it to end-use customers is complex; varies by locality, state, and region; and involves many different actors and interests. Likewise, how demand response is used today also varies greatly by locality, state, and region. For example, some demand response is utilized through regional wholesale markets and programs, and some demand response is operated at the retail electricity level, through programs that are under the jurisdiction of and overseen by state regulators or locally appointed or elected governing boards for publicly and cooperatively owned electric utilities. These differences should be acknowledged so that the framework of the National Action Plan is inclusive and can be tailored to each locality, state, and region.

Encouraging the participation of local, state, and federal government actors as well as utilities/load-serving entities, demand response providers, RTOs/ISOs, and customers (industrial, commercial, and residential customers) will help ensure that the National Action Plan reflects regional differences and taps into the expertise and experience of various groups. A Coalition will give these stakeholders an opportunity to work together to build support for a comprehensive national strategy and ensure its success.

Who Should Be Involved?

For the effort to be a success, the Coalition should be composed of volunteers from:

- Federal, state, and local governments
- Utilities, load-serving entities, and other demand response providers
- RTOs/ISOs
- Commercial and industrial customers
- Consumer advocates
- Existing coalitions
- Interest groups and other stakeholders

What Elements Make a Coalition Successful?

There are many factors that help a Coalition succeed:

1. **Clear goals.** The Coalition stakeholders need to have a clear understanding and agreement on the goals they are trying to meet.
2. **A clear organizational structure.** Stakeholders of the Coalition need to understand how decisions are made, how work will be completed, and how accountability for achieving the goals of the National Action Plan will be measured.
3. **Good leadership.** An effective Coalition is composed of effective volunteers and, depending on the nature of the work, strong guides who provide assistance and direction to help the volunteers achieve their goals.
   a. Effective volunteers. First and foremost, stakeholders should be cooperative and support the National Action Plan goals. These stakeholders should be leaders who are empowered to make decisions for their respective organizations. They also must have time to commit to the responsibilities of the Coalition.[31]
   b. Effective volunteer leaders. A smaller subset of stakeholders who are well- versed in working with other volunteers and are willing to provide leadership to the overall Coalition.
   c. Assistance and support. Given that stakeholders will be volunteering their time, it may be important to have an entity or a federal agency (for example, FERC and/or DOE) observe the progress of both federal and private efforts to complete the National Action Plan activities, ease implementation by tracking these efforts, be aware of any gaps in implementation, and, where appropriate, encourage and work with federal and/or private actors to fill the gaps.
4. **Funding.** Each stakeholder will fund its own participation in the Coalition.

Under the National Action Plan, the Coalition would be established with the participation of all stakeholders. The Coalition could be established under a number of different scenarios. For example, part of the National Forum could be dedicated to identifying an organizational structure for the Coalition. Whatever means is chosen to establish a Coalition, stakeholders should consider leveraging the experience of existing coalitions to determine an appropriate format for collaborating on the activities identified in the National Action Plan.

How Should the Coalition Operate?

The Coalition would serve as a forum for its stakeholders to coordinate implementation of the National Action Plan. While each stakeholder in the Coalition would be responsible for implementing particular activities identified in the National Action Plan, the Coalition would provide an opportunity for these entities to share insights, experiences, current gaps, and areas of mutual concern or need and agree on coordinated strategies.

Under the National Action Plan, one or more federal agencies participating in the Coalition (for example, DOE where appropriate) would direct and perhaps also fund implementation of certain identified activities of the National Action Plan that it chooses to fund. Activities identified in the National Action Plan that lend themselves to being accomplished by private entities would be accomplished using private funding. Existing coalitions could share their expertise and may be able to provide long-term support and leadership for the Coalition. The work of these private entities would be coordinated with that of federal agencies through the Coalition to the extent practical and legally permitted. This would help ensure that the efforts of governing officials, utilities/load-serving entities, RTOs/ISOs, demand response providers, consumer advocates, commercial/industrial customers, and other stakeholders are leveraged and/or enhanced.

In addition, it also could be useful for a federal agency (for example, FERC and/or DOE) to observe the progress of both federal and private efforts to complete the National Action Plan activities, ease implementation by tracking these efforts, be aware of any gaps in implementation and, where appropriate, encourage and work with federal and/or private actors to fill the gaps.

1.2. Examples of the Coalition Approach

The use of a Coalition has been successful in similar national efforts. Two examples, differing in objectives and approach, illustrate the Coalition approach to communicating the benefits and opportunities for demand response and to developing the methods and tools that states, utilities, demand response providers, and others involved find useful and beneficial to consumers. Appendix A contains more information on these two examples.

The National Action Plan for Energy Efficiency

In 2006, a Leadership Group composed of more than 50 leading electric and gas utilities, state utility commissioners, state air and energy agencies, energy service providers, energy consumers, and energy efficiency and consumer advocates released the *National Action Plan for Energy Efficiency*.[32] This report is a call to action to increase investment in ratepayer- funded energy efficiency; it lays out five main policy recommendations to achieve greater investment from sponsoring organizations. The Leadership Group then called for development of a set of resource materials, largely best practices based on the last two decades of experiences in implementing ratepayer-funded energy efficiency, for parties who are interested in pursuing ratepayer-financed energy efficiency. In 2008, the *National Action Plan for Energy Efficiency* was updated with the *Vision for 2025*,[33] which identifies 10 implementation goals and measures the progress made so far. Recently, the Leadership Group sponsored a report to explore the coordination of demand response and energy efficiency, which is described in Sidebar 2.

The initial Leadership Group was co-chaired by the then-president of the National Association of Regulatory Utility Commissioners (NARUC), Diane Munns, and the President and Chief Executive Officer of Duke Energy, James Rogers. DOE and U.S. Environmental Protection Agency (EPA) participation in the Leadership Group is limited to facilitation. As of January 2010, more than 120 organizations have endorsed the *National Action Plan for Energy Efficiency* recommendations and/or made commitments to promote energy efficiency within their spheres of influence. The Leadership Group and its National Action Plan for Energy Efficiency ended its activities in November 2009, after having achieved its original purpose of serving to draw member attention to energy efficiency.

The Air Bag and Seat Belt Safety Campaign

In 1993, General Motors and the National Safety Council formed the Air Bag and Seat Belt Safety Campaign, a coalition of automobile manufacturers, insurance companies, the Insurance Institute for Highway Safety, safety organizations, law enforcement associations, and the National Highway Traffic Safety Administration, to address the issue of child air bag deaths and stagnant nationwide seat belt use.[34] They conducted a national umbrella campaign to encourage parents to restrain their children properly in the back seat, and to encourage all drivers and passengers to buckle up. This campaign also provided media and organizational outreach materials to states and communities so that they were well prepared to serve as spokespeople for the effort even if they were not demonstration program states.

As a result of the Air Bag and Seat Belt Safety Campaign's work in demonstration states and then nationwide, seat belt use increased more than it ever had during a prior period, from 61 percent in 1996 to a record high of 82 percent in 2005.[35] Child fatalities from traffic crashes dropped by 20 percent, and proper restraint use among toddlers jumped dramatically from 60 to 94 percent.[36]

SIDEBAR 2: COORDINATION OF ENERGY EFFICIENCY AND DEMAND RESPONSE

In support of the 10 implementation goals outlined in the *National Action Plan for Energy Efficiency's Vision for 2025*,[1] the National Action Plan for Energy Efficiency, with input from the Leadership Group, released a paper discussing the coordination of energy efficiency and demand response in January 2010.[2] The paper defines energy efficiency as "using less energy to provide the same or improved level of service to the energy consumer in an economically efficient way."[3] On the other hand, demand response "entails customers changing their normal consumption patterns in response to changes in the price of energy over time or to incentive payments designed to induce lower electricity use when prices are high or system reliability is in jeopardy."[4] The paper also recognizes that energy efficiency and demand response currently have significant differences in how they are measured, what organizations offer them, how they are delivered to customers, and how they are rewarded in the marketplace. In order to reduce these differences and achieve possible benefits, the paper outlines ways in which energy efficiency and demand response can be coordinated. It discusses the concerns and barriers affecting the coordination of energy efficiency and demand response, and articulates steps that can be taken to achieve better coordination.

Energy efficiency and demand response programs are rarely combined into an integrated program.

The paper states that "in December 2009, out of 2,016 U.S. and Canadian energy efficiency, demand response, and load management programs in the E Source database, only 56 were identified as serving both energy efficiency and demand response purposes."[5]

Therefore, one way in which to better coordinate energy efficiency and demand response is to offer combined offerings in the future. Even if energy efficiency and demand response programs are not combined, program sponsors could provide marketing and education on these programs in a more unified way. Furthermore, both private firms, via finding customers interested in reducing their energy costs, and the government, via building codes and appliance standards, can contribute to the simultaneous achievement of energy efficiency and demand response goals.

Nevertheless, the integration of energy efficiency and demand response programs remains a difficult and complex endeavor for several reasons. First, the funding for energy efficiency and demand response can sometimes come from different sources. Second, some energy efficiency managers have expressed reservations about demand response interfering with their programmatic goals. Third, retail rates need to be designed with energy efficiency and demand response objectives in mind — i.e., dynamic or time-varying rates may be best offered with inclining block rates. Finally, it will take time to develop the capabilities of utility staff and contractors in both energy efficiency and demand response.

The paper states that many actions can be taken in order to encourage better coordination of energy efficiency and demand response. Examples include clear regulatory signals that encourage utilities and grid operators to coordinate energy efficiency and demand response activities, better articulated demand-side management program goals that target both energy savings (energy efficiency) and peak demand reductions (demand response), and a new set of consumer education activities that simultaneously explicate energy efficiency and demand response benefits.

[1] National Action Plan for Energy Efficiency, supra note 28.

[2] National Action Plan for Energy Efficiency, Coordination of Energy Efficiency and Demand Response (2010) (Prepared by Charles Goldman (Lawrence Berkeley National Laboratory), Michael Reid (E Source), Roger Levy, and Alison Silverstein), available at www.epa.gov/ eeactionplan.

[3] *Id.* at ES-1.

[4] *Id.*

[5] *Id.* at ES-3.

PART 2: STRATEGIES AND ACTIVITIES

This section sets out a number of strategies and activities that will be pursued through the National Action Plan to fulfill the objectives in EISA section 529.

The activities are organized into three categories: technical assistance to states, the national communications program, and development or identification of tools and materials for use by customers, states, and demand response providers. However, there will be significant, inherent overlap between these categories of activity. Requirements for tools and materials can be developed and coordinated at the same time as requirements for technical assistance and the national communications plan. Similarly, providing messages and talking points to states that implement demand response would be a component both of the communications program and of technical assistance to states. For the sake of simplicity, action items are listed under the statutory objective areas that are most closely related to each activity. Figure 1 illustrates the functional overlap between the three activity areas.

Given the overlap in technical assistance and tools and materials, we anticipate that not only states but also customers, utilities, demand response providers, and other bodies and institutions will be able to take advantage of the resources identified in the National Action Plan. It should be noted that the degree to which all elements of the National Action Plan are developed is heavily dependent on the level of funding and organization that the Coalition achieves. The options presented in the following sections could be pursued more or less aggressively than as they are currently described, depending on how these factors unfold in the implementation phase of the National Action Plan.

To meet the three objectives in EISA, the National Action Plan will be applicable to a wide range of demand response programs, products, and technologies, and it also will be applicable to various stages of development and deployment of demand response implementation.

2.1. Technical Assistance to States

Introduction

Section 529 of EISA requires that the National Action Plan identify what technical assistance[37] states need to maximize the amount of demand response resources that can be developed and deployed. For the purposes of this

National Action Plan, targets for state assistance include governors, state regulators, state energy offices, state legislators, and state consumer advocates. While the statute requires the National Action Plan to identify "the requirements for technical assistance to States," demand response activities of publicly owned and cooperatively owned utilities do not generally fall under state commission jurisdiction. Thus, providing technical assistance only to state agencies will not achieve the goals of providing assistance to all governing bodies responsible for overseeing the provision of electric service to all customers. Local officials governing publicly owned and cooperatively owned utilities face challenges similar to those of state governing officials. Therefore, this section's discussion of most types of technical assistance is geared toward the needs of state and local governing officials.

In many states, a variety of entities are already providing demand response resources. (See Appendix B for examples.) These states tend to be those that have passed laws and/or regulatory mandates or action plans that incorporate demand response in the energy policy formation process. Often, these states (for example, Connecticut) have laid out a framework for integrated resource planning that explicitly incorporates demand response as a resource.[38] In some states, RTOs and ISOs operate organized energy and capacity markets into which demand response can be bid as a resource. Appendix B also lists current opportunities for demand response to participate in RTO/ISO markets.

Several states have taken regulatory or legislative actions to encourage demand response and energy efficiency.[39] For example, the Michigan Public Service Commission's *Michigan's 21st Century Energy Plan*[40] and the New Jersey Board of Public Utilities' *Energy Master Plan*[41] set forth targets to reduce energy consumption and peak demand for the utilities and/or load-serving entities operating in those states. Massachusetts' Green Communities Act,[42]

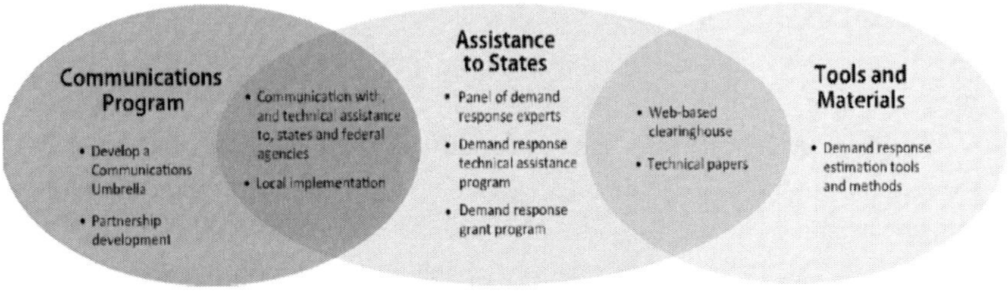

Figure 1. Functional Overlap in Some National Action Plan Activities.

Maryland's EmPOWER Maryland Energy Efficiency Act of 2009,[43] and Pennsylvania's Act 129[44] set goals for reductions in energy consumption and peak demand. In Colorado, the legislature recently passed legislation that permits cooperative electric associations to introduce inclining block rates for residential customers to promote energy efficiency.[45] In Iowa, utilities are required to develop energy efficiency programs,[46] which include demand response.[47]

Many state and local governing officials and other key stakeholders (such as other demand response providers and public power entities, including rural cooperatives and publicly owned utilities) recognize the potential benefits of demand response programs, products, technologies, and incentives and would like to increase participation and expand their deployment.[48] However, as discussed at the FERC staff technical conference on November 19–20, 2009, inadequate training, time constraints, and staffing or resource limitations can impede effective implementation and deployment of demand response programs. While wholesale markets are seeing increasing participation by providers of demand response resources, customers who participate in these programs are primarily in the commercial and industrial sectors and represent a relatively small fraction of all retail customers.

Several organizations, such as load-serving entities, RTOs/ISOs, and other demand response providers, are developing ways to enable demand response to be utilized more broadly as a resource in energy, capacity, and ancillary services markets. These novel approaches will also help foster the full deployment of AMI and dynamic pricing. In addition, they will release competitive market forces to 1) develop more customer demand response and 2) plan for the deployment of many innovative demand response applications that create greater consumer control over energy usage and create new cost-saving opportunities for consumers.

Technical Assistance Actions

Section 529 of EISA requires that the National Action Plan identify the requirements for technical assistance to states. Assistance to other entities — such as utility and non-utility demand response providers and consumers — is discussed in Sections 2.2 and 2.3 of this document. The overall goal of technical assistance is to provide high-quality information and research to states. States will use this information to make decisions regarding demand response policy, regulation, and investments. Technical assistance can be provided by private entities or by federal agencies. For instance, DOE currently gives states assistance that is related to demand response as one of its many state electricity topics.

Establish a National Forum

Purpose	Give federal, state, and local leaders, as well as key stakeholders, a venue to introduce the National Action Plan.
Expected Outcome	Awareness of and support from policymakers for accomplishing the activities identified in the National Action Plan.

As the first step in providing technical assistance under the National Action Plan, a federal agency will sponsor a national forum on demand response that would be held for federal agencies, state public utility commissioners, state energy offices, gubernatorial offices, state legislators, state consumer counsels, utilities, other key stakeholders, and interested consumers. Such a forum would provide a useful opportunity for the Coalition, to present the overall vision of the National Action Plan to key decision-makers on demand response. The national forum also would provide an opportunity to share ideas, examine barriers, and explore solutions. Part of the National Forum could also be dedicated to developing an organizational structure and leadership for the Coalition.[49]

The forum would be a stand-alone, one-day session, likely in Washington, D.C., or following another national conference (e.g., a meeting of NARUC). The forum also could be integrated with existing third-party forums and meetings on demand response to leverage existing infrastructure, marketing, and name recognition to attract broad participation. The national forum also could take a blended approach that allows for in-person participation as well as online participation, e.g., webcasting to reach the widest possible audience.

After the initial national forum, the Coalition would promote and develop a series of annual meetings. The purpose of these meetings should be to review results that have been achieved, share information on the latest developments in demand response, and continue the provision of targeted assistance to states as needed. If annual meetings are determined to be useful, the locations of these national meetings could rotate between regions (e.g., East Coast, Midwest, West Coast) to foster greater in-person participation. If the forum is held once per year, the ideal time would be in the fall, when the results of summer demand response efforts would be known. The forum would run no longer than two days. In addition, to maximize participation, these meetings should be designed as webinars to encourage broad participation.

Conduct Informational and Educational Sessions for Policymakers and Regulators

Purpose	For state and federal regulators, legislators, and other policymakers to share and learn about the attributes, uses, costs, and benefits of effective demand response and opportunities to implement demand response.
Expected Outcome	State and local governing officials are better prepared to implement demand response.

Following the national forum, multiple regional and state sessions should be held, targeted to a broader set of state employees, regulators, and other stakeholders. Regional and state sessions may be more convenient for other interested retail regulators such as city councils and rural cooperative boards. The objectives of the regional sessions would include those of the national forum but also should seek to coordinate and implement applicable activities proposed in the National Action Plan for the deployment of demand response in the region. In addition, these regional and state sessions would be coordinated with existing regional dialogues (e.g., efforts in the Midwest, Mid-Atlantic, and Pacific Northwest) funded by DOE at the request of the states.

Regional sessions would allow more targeted discussion and input from state officials and local utilities on demand response issues specific to each region — for example, demand resource participation in wholesale organized markets in regions served by RTOs or ISOs, and use of demand response to meet ancillary service needs associated with integration of renewables. The sessions could address the various services provided by demand response and the associated costs and potential benefits of demand response programs, products, technologies, and incentives to the states and their electric customers. These sessions also could help stakeholders develop strategies for deployment.

It is expected that the informational and educational sessions should promote knowledge and skills related to:

- The various services provided to enable demand response.
- The associated costs and potential benefits of demand response programs.
- Products, technologies, and incentives available to states and their electric customers.

- Strategies for deployment.
- Practical aspects of demand response implementation.
- Just-in-time results and best practices from demand response implementation.
- Relationship of demand response to the smart grid and the deployment of smart meters.
- Navigation of federal/state jurisdictional issues related to demand response.

The Coalition could sponsor these regional sessions and recruit speakers and facilitators for these sessions. In addition to conducting the workshops, these speakers and facilitators should be responsible for developing workshop materials and customizing these materials to address specific regional and state issues.

Examples: Informational Sessions

CPUC and California Institute for Energy and Environment

CPUC and the California Institute for Energy and Environment have developed a series of white papers on various behavioral topics and conducted two-hour webinars to present the papers and discuss the results. CPUC funds the working papers and the Institute selects the experts to prepare and present the papers. Papers and presentations are then posted to the Institute's website.

http://uc-ciee.org/energyeff/energyeff.html

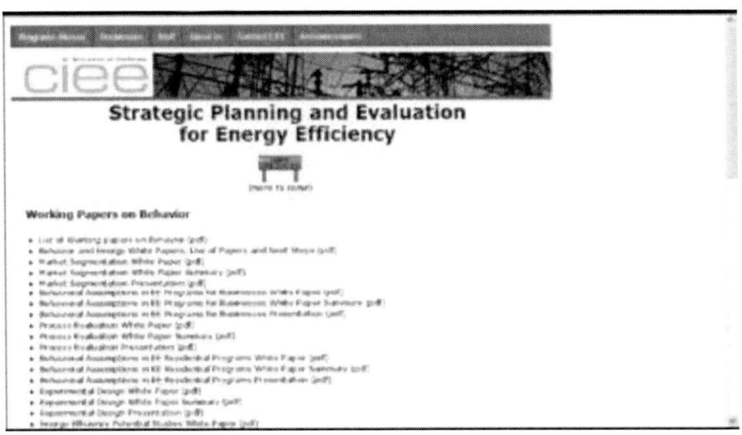

National Regulatory Research Institute

The National Regulatory Research Institute provides education sessions for various topics, such as smart grid issues and dynamic pricing. Slide decks and information related to these presentations are posted to their website.

http://www.nrri2.org/

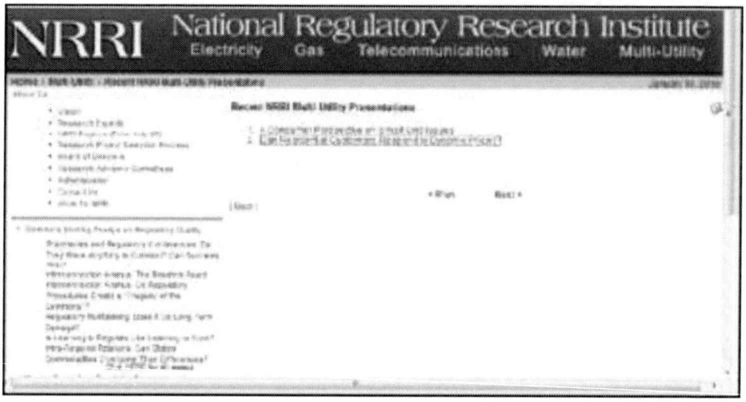

Build a Panel of Demand Response Experts

Purpose	Deliver lectures on and generate interest in demand response social and technical topics at national and regional conferences; consult on demand response programs and projects.
Expected Outcome	Enhanced knowledge, skills, and attitudes of stakeholders regarding demand response programs; technologies; markets; statutes; and/or ordinances, benefits, and economics.

The Coalition would oversee the recruitment of experts to help inform various constituencies about demand response programs, products, technologies, and incentives and their benefits. These experts could be available as contacts for state and local governing officials during policy development or regulatory proceedings.[50]

The Coalition would establish an inventory of demand response industry technical experts. One means to manage this panel would be through a Web-searchable database through which states and other entities could identify a short list of experts for a specific topic. For example, a user could use this database to find someone with expertise related to customer education for small/medium commercial entities, or to find someone with expertise related to automated demand response in aluminum manufacturing facilities.

The initial panel of experts could be developed by soliciting recommendations from the federal agency, as well as Coalition members, and by identifying experts who have spoken at industry conferences and published in relevant journals. Additional experts could then be identified through recommendations from existing panelists. All experts would need to disclose conflicts of interest, which is a best practice found in health-care-oriented conferences and coalitions such as the American Public Health Association. To encourage freedom of expression and diversity of thought, the messages communicated by experts should flow freely. Therefore, the views presented by these experts should not be construed to represent the opinion or policy of the federal agency or the Coalition.

The panel of experts could serve as a resource for state and local government officials as they consider demand response programs. The experts should provide information about the types of demand response programs and assist state and local government officials in assessing demand response. In addition, experts could bc speakers at events. The Coalition could provide speakers through both "push" and "pull" methods. Push methods involve Coalition members proactively recommending a panelist to participate in an event. Pull methods reflect states or entities in need of panelists for events searching the panelist database themselves to identify experts for speaking engagements, or asking the Coalition or federal agency to recommend an appropriate expert. Regardless of the method, the entity sponsoring an event in which a panelist will participate would be responsible for selecting the appropriate panelists for their event.

Sponsor Technical Papers

Purpose	Generate new knowledge or synthesize existing knowledge to inform state officials about the design of demand response programs, products, technologies, incentives, markets, and legislation.
Expected Outcome	Peer-reviewed or similar-quality papers and their distribution through an online database with clear summaries/abstracts.

Technical knowledge about demand response is underdeveloped, underfunded, limited in scope, decentralized, disorganized, and in some cases proprietary and thus hidden from public view. There exists no peer-reviewed journal for demand response research, no online database of demand response research citations, and no shared agenda and prioritization for ongoing demand response research. Published research is generally limited to trade press (such as *The Electricity Journal and Public Utilities Fortnightly*), reports independently developed by utilities and research organizations, and peer-reviewed journals that are not specifically dedicated to demand-side issues (such as *Energy, Energy Economics, Energy Policy, The Energy Journal, The Journal of Regulatory Economics, and Utilities Policy*). Some state organizations such as the California Energy Commission's Research, Development, and Demonstration Division (formerly the Public Interest Energy Research program) set research agendas, solicit proposals, and award funding (collected from public interest funds) for research.

To promote the development of technical knowledge about demand response, the Coalition would solicit information about the demand response topics that would possibly be helpful to state officials and would oversee the development of a series of informational, technical, research, or policy papers targeted to various practical aspects of demand response implementation for the use of state officials. These papers could be disseminated through the clearinghouse discussed in Section 2.3, though if the federal agency sponsors the clearinghouse, various standards on objectivity and public interest filters would be applied. Overall, the primary purpose of the technical papers would be to highlight and address questions that require new research to address barriers to demand response.

The Coalition and its stakeholders would develop and prioritize topics of interest. The Coalition and its stakeholders could solicit research proposals from national laboratories, academic institutions, utilities, independent researchers, consultants, and other entities. Subject to available funding, the Coalition or the federal agency may be able to award funding to proposals that have merit. If the federal government sponsors any technical papers identified by the Coalition, attention will focus on avoiding conflicts of interest, because members of the Coalition may have financial interests in the message developed. Proposal solicitation should embrace both primary (original) research and secondary (meta-analysis-type) research, with an emphasis on the latter to bring forward existing proprietary research commissioned by utilities. There also could be crossover between a demand response research agenda and an energy efficiency research agenda (for example, in the social-behavioral area), so there should be an attempt to align research with these other areas and to coordinate with existing research efforts. For example, a number of utilities receiving smart grid investment grants from DOE have proposed consumer behavior studies as part of their smart grid/demand response rollouts.

Some potential research topics that may assist states in addressing information gaps are identified in the discussion of the national communications program, and some are listed in Appendix A, where they are divided into two categories: social science[51] research issues or technical issues. The Coalition and its stakeholders would work to prioritize these topics, identify additional topics, and select a subset as candidates for further research.

Establish a Program for Technical Assistance for Demand Response

Purpose	Provide support to states for implementing demand response.
Expected Outcome	Demand response expertise is provided to states requesting assistance in the development and deployment of any and all forms of demand response.

The provision of research and technical support in response to state requests is a core activity in any effort to provide technical assistance to states. DOE has provided federal support for this type of technical assistance for many years. As part of their support, DOE supports research and technical assistance activities to ISOs, utilities, and state and regional policymakers that support and enhance the development of demand response programs, technologies, services infrastructure, and dynamic pricing tariffs.

This activity involves establishing a program under which qualifying state agencies (e.g., state regulatory commissions or state energy offices), consumer advocacy groups, and utilities may apply for specific virtual or onsite technical assistance with demand response implementation. Subject to available funding, the federal government or the Coalition could oversee the provision of appropriate technical assistance to stakeholders. The nature of this assistance could range from referring a caller to a website, to providing the applicable reference material, to providing onsite technical help (for a day, a week, or longer). This program could also provide assistance to help states navigate federal/state jurisdictional issues related to demand response. The program would be coordinated with ongoing DOE technical assistance programs.

The technical help could be provided through a number of avenues, including the following:

- Staff of a federal office.
- National laboratory staff.

- Entities specializing in operational and dispatch algorithms to allow demand response to be incorporated as a resource, similar to generation resources.
- Funding for lead staff from a state that has successfully implemented a particular demand response program to travel to, and inform, another state developing a similar program.
- Entities specializing in demand response communications and marketing.
- Support for the NARUC-FERC Smart Response Collaborative.
- Other knowledgeable individuals.

Provide Demand Response Grants

Purpose	Provide funding for demand response innovations (subject to available funding).
Expected Outcome	Innovative solutions to demand response problems.

Some states provide grants through regulatory entities — for example, the California Energy Commission's Research, Development, and Demonstration Division (which administers $83.5 million in public interest energy research funds). The California grant program focuses on developing and implementing commercially available demand response programs. In addition, DOE is using American Reinvestment and Recovery Act (Recovery Act)[52] funds to support the deployment of many demand response technologies, programs, pilots, and demonstrations, as well as studies of customer behavior response. These grants will entail very extensive data collection and reporting, available publicly on the Internet through www.smartgrid.gov, and will provide a wealth of information that state regulatory authorities can use to make future decisions about demand response-related investments.

The purpose of providing demand response grants would be to fund additional and targeted development of demand response at the state level through pilots, implementation, and evaluation of demand response innovations. This evaluation would be designed to build on and leverage the activities funded at the state level and through the Recovery Act. Because innovations can emerge from a variety of sources, states or other legal entities, public or private, should be eligible to apply for a grant. Also, many states are not structured for distributing grants, so another entity would need to provide funding and oversight of the grant review process. Reviewers could evaluate and score grants based on criteria established by the Coalition. For instance, the criteria could favor programs, products, services, and technologies that have the potential for maximum, sustained deployment of cost-effective demand response. Other criteria could include significant potential as established by scientific evidence, work in untested areas with strong potential, compliance with existing standards, and novel demand response applications that address climate change issues. Once grant winners are announced, an entity that has an existing grants administration infrastructure, working in conjunction with the Coalition, would oversee the relationship with the grantee, from award to final report.

Examples: Grant Programs

DOE's SBIR/STTR Program

Each year (typically around the beginning of October), DOE issues a solicitation inviting small businesses to apply for SBIR/STTR Phase I grants. This program covers technical topics in such research areas as energy production (fossil, nuclear, renewable, and fusion energy), energy use (in buildings, vehicles, and industry), fundamental energy sciences (materials, life, environmental, and computational sciences and nuclear and high energy physics), environmental management, and nuclear nonproliferation. Grant applications submitted by small businesses must respond to a specific topic and subtopic during an open solicitation.

http://www.er.doe.gov/sbir/

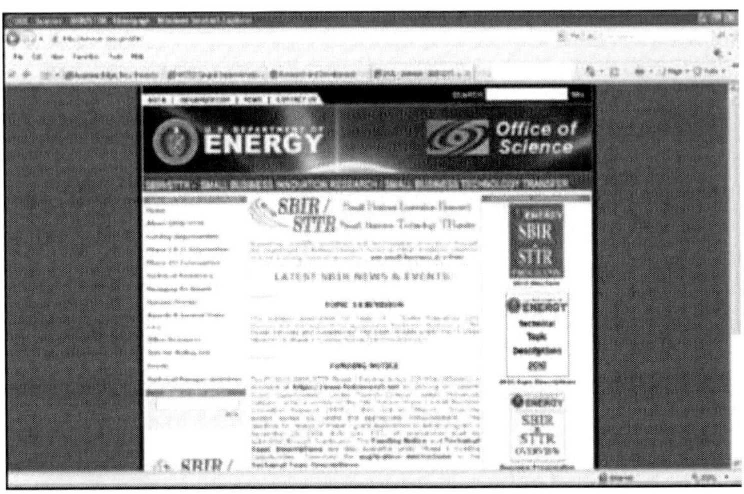

California Energy Commission

The California Energy Commission's Research, Development, and Demonstration (RD&D) Division supports public interest energy research, development, and demonstration. RD&D activities include providing contracts and grants for research and development of energy technologies and related scientific activities.

The RD&D Division administers a total of $83.5 million in public interest energy research funds annually — $62.5 million for electricity and $21 million for natural gas.

http://www.energy.ca.gov/research/index.html

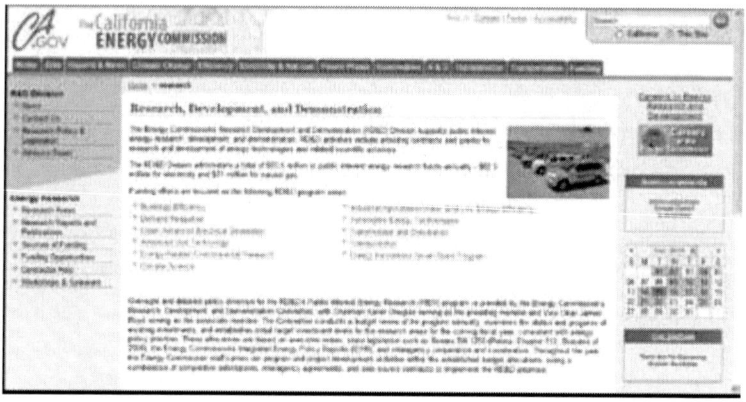

2.2. National Communications Program

Introduction

Section 529 of EISA requires that the National Action Plan on Demand Response design and identify requirements for implementation of a national communications program that includes broad-based customer education and support. This section describes the design of, and identifies the requirements for, a national communications program that includes three elements: a Communications Umbrella; local implementation; and direct outreach to states, policymakers, and national partners regarding demand response communications programs and opportunities.

A national communications program should be multifaceted to facilitate widespread acceptance and implementation of demand response to achieve the highest potential peak load reduction identified in the National Assessment. To account for the various approaches undertaken and considered for demand response implementation by states and localities, a national communications program would not be a single, national effort to educate all consumers at the same time, through a single channel and with the same message; rather, the national communications program would take a more tiered, localized approach that educates customers, as appropriate to each locality. A national communications program also must be highly adaptable and scalable to individual states, utilities, load-serving entities, and other demand response providers, without taking a top-down or one-size-fits-all approach. Simultaneously, the program would benefit from the coordinated research and insight that only a nationwide program can provide. It would not dictate a single message, which may not apply to all customers or electricity stakeholders; instead, it would provide a framework of messages that universally appeals to customers, but that can be employed in part or whole — as deemed appropriate by local demand response implementers — to each region, state, or locality.

Accordingly, this national communications plan provides the roadmap for a communications program that will educate residential, commercial, and industrial electricity customers, and support those customers as well as electricity providers, regulators, policymakers, and all others involved in implementing demand response programs. Designing the plan as a flexible roadmap for the national communications program enables communications around a variety of applications. These applications include incorporating price-responsive demand, automating demand response in all hours of the day, using demand response to provide operating reserves and other ancillary services, and partnering demand response with wind generation and other variable renewable resources.

This plan for a national communications program reflects the input of a broad range of industry stakeholders received through extensive outreach, a staff technical conference, and formally filed comments in response to the Discussion Draft and the Draft National Action Plan.

The multifaceted structure of the national communications program would mean that it would develop and support communications through multiple channels and media to communicate as appropriate with the many audiences for demand response. The national communications program would provide broad-based customer education and support though coordinated national and local action. At the national level, the program would develop a Communications Umbrella to provide the research-based message framework, communications tools, support, and advice, as requested, to local demand response implementers.

In addition, the Coalition would oversee direct outreach and educational programs at the national level to states, policymakers, and national partners to build support for demand response. By communicating with states, policymakers, and national partners, the national communications program would help build support for demand response in areas where there are no existing demand response programs or current plans to implement such programs.

At the local level, the national communications program should include an implementation strategy to inform residential and commercial customers about the availability and possible benefits of demand response programs, products, technologies, and incentives. Interested local demand response implementers would use the research-based message framework and communications tools and tactics developed at the national level to tailor messages and materials to conduct their own outreach to their particular customers. Local implementation includes demand response provided in regional RTO/ISO markets, states, and localities. The local implementation strategy would have two components: a demonstration program approach and support for other demand response programs. Under the demonstration program approach, local demand response implementers could volunteer to participate as demonstration programs to receive "hands-on" implementation message support and tactical advice. The demonstration programs could thereafter serve as successful program models or case studies that could then further inform the national message framework.

The other component of local implementation should be support for other demand response programs. Local demand response implementers would not need to participate as demonstration programs to utilize resources developed at the national level. Local stakeholders who wish to take a more autonomous approach to demand response implementation, using only certain communications tools or resources as they determine appropriate, would fall into this category but would otherwise have access to the same message framework and tactical implementation recommendations as would demonstration programs. All local demand response implementers could access available toolkit materials to supplement their demand response efforts. As discussed later in this section, these materials would be customizable so that implementers could modify and tailor them to meet their particular needs and achieve their specific objectives. By targeting the program at the local, regional, and state levels, the communications messages would be more effective, and feedback on what works best could be used to influence the national effort.

Figure 2 shows the proposed national communications program structure, outlining the program and message development tasks to be overseen by the Coalition in the Communications Umbrella. The research and messages developed in the Communications Umbrella would then be used to 1) support local demand response implementers involved in demonstration programs and other demand response programs and 2) reach out to states, policymakers, and national partners to educate them about demand response. Later in this section, the structure and elements of the national communications program are discussed more extensively.

Figure 3 shows the approximate sequence in which major tasks would be undertaken, indicating the overlap between some tasks and the ongoing nature of many. The diagram indicates that certain tasks, such as communications with states and policymakers, would occur over a long duration, while other tasks, such as foundational customer research and message development, would be more limited in duration in the initial stages (though possibly repeated at a later time if warranted). Multiple tasks would occur concurrently based both on their priority level and on the steps that must occur prior to each stage of the rollout.

Challenges in Designing a National Communications Program

The national communications program could face several challenges. First, no single nationwide demand response program is currently in place. Many different demand response programs, products, technologies, and incentives are being considered or implemented. Customer knowledge pertaining to demand response also varies greatly. In addition, because demand response implementation varies depending on the circumstances faced in various regions, it may be difficult and inappropriate to develop a "one-size-fits-all" message.[53] Other challenges that a national communications program must consider include the diversity of stakeholder interests, the lack of broad-based knowledge about customer perceptions and motivations, the various energy initiatives competing for customers' attention and funding, and the degree of receptivity to demand response and technical assistance among local entities. A national mass media campaign directly targeting residential and small commercial customers could be undertaken, depending on the availability of financial resource.[54]

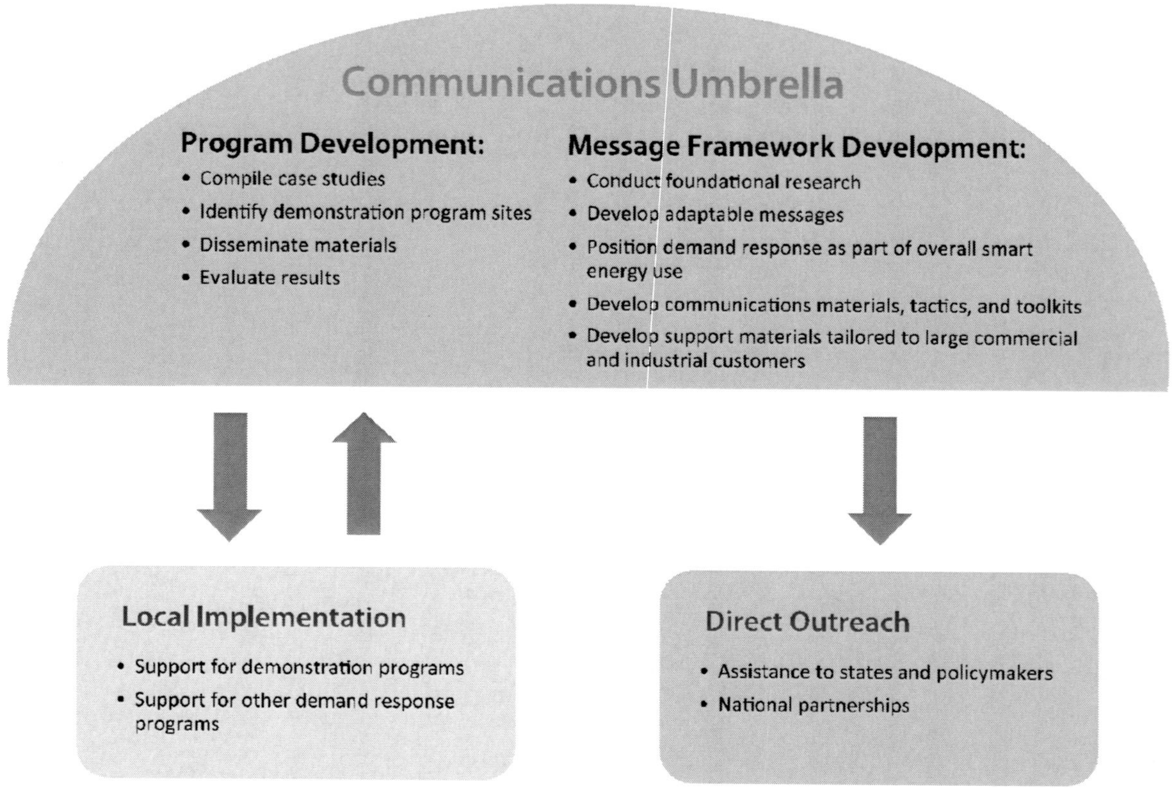

Figure 2. Program Structure.

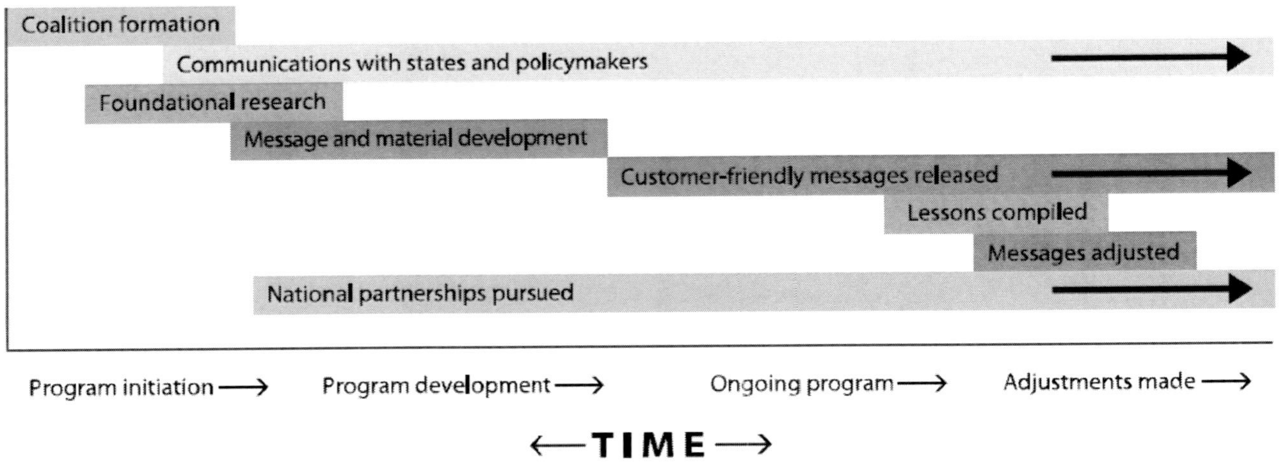

Figure 3. Program Development and Rollout Sequence over Time.

Based on these challenges, the Coalition should initially focus on supporting local implementation efforts through national research and coordination rather than by initiating a national mass media campaign. By supporting local efforts, the program would ensure that resources are spent to reach customers only when demand response programs are available to them. This feature helps ensure that a national communications program does not expose customers to confusing or conflicting messages, because local utilities, load-serving entities, other demand response providers, governors, state energy offices, and public utility commissions would serve as the chief communicators, and new messages would only supplement their current communications, not supplant them. By reaching out only to customers who, according to research, are receptive and can act on the information, the program will be more likely to be effective. However, this discussion should not been seen as limiting future efforts to reach more customers using the lessons learned from the National Action Plan effort.

Objectives of the National Communications Program

With these challenges in mind, the national communications program has the following objectives:

- Increase electricity customers' awareness and understanding of their energy-consuming behavior and the means to adjust their consumption to realize the most economical use of electricity.
- Increase commercial and residential electricity customers' awareness and understanding of demand response as it relates to other smart energy use principles such as energy conservation, energy efficiency, and the smart grid. (Research and outreach have shown that many large industrial customers are already largely aware of the potential benefits of demand response as well as how to implement demand response initiatives.)
- Provide information to relevant states and policymakers (including, but not limited to, state and local governing officials, such as state legislators) about the potential benefits and costs associated with demand response, including peak reduction and ancillary services to reliably integrate variable resources.
- Inform policymakers about the potential benefits of having a demand response program that could complement their smart grid, renewable energy, and environmental goals.

- Inform renewable resource developers of the benefits of "partnering" with demand response to improve reliability and provide needed ancillary services.
- Support local utility, load-serving entity and other demand response providers' activities that inform customers about the potential opportunities, benefits, and costs of demand response and thereby engender customer support and participation in demand response in their local programs.
- Establish measurable goals to meet the national communication program's objective of broad-based customer education and support. The development of these goals would be coordinated by the Coalition, with benchmarks to facilitate program evaluation and accountability.
- Address concerns about negative effects on low-income, elderly, and disabled customers.

National Communications Program Structure

The national communications program would function as a forum for communications development and as a supporter and resource for regional and local outreach to residential, commercial, and industrial customers. By compiling research on existing communications and marketing programs across the country and internationally, and conducting foundational research among electricity customers, the Communications Umbrella approach would be used to develop a message framework that local utilities, load-serving entities, and other demand response providers could employ at the local level. It would not take a top-down approach. Rather, it would look to the experience and best practices of local demand response implementers as the foundation of the Communications Umbrella approach, and conduct research only after understanding customers' existing base of knowledge. Foundational customer research and analysis of the existing body of research would then proceed concurrently.

This national communications plan outlines the specific roadmap for the national communications program that includes broad-based customer education and support. This program would have the following structural elements:

Communications Umbrella

- Program development
- Message framework development

Local Implementation

- Support for demonstration programs
- Support for other demand response programs

Direct Outreach

- Communications assistance to states and policymakers
- Partnership development

Each of these elements will be discussed in further detail below.

Communications Umbrella

Purpose	Provide a consistent communications program message framework, based on research conducted as part of the National Action Plan and additional, publicly available research and case studies that can be tailored to meet the unique needs of localities, states, and regions. Provide a well-researched strategic framework for communicating about demand response in localities, states, and regions.
Expected Outcome	A body of customer research, a clear message framework, customizable message and communications materials, and a strategic communications roadmap for demand response implementation.

A Communications Umbrella would provide a foundation for successful future communications about demand response, by leading the way on market research, creating a message framework, and providing outreach strategies and tactics on which to base national, regional, state, and local communications efforts. Figure 4 illustrates this relationship.

The Communications Umbrella would create a consistent message framework that local demand response implementers (especially utilities, load-serving entities, and other demand response providers) can customize and use to develop their own message and materials. More specifically, the Communications Umbrella would:

1. Provide communications support and advice to local demand response implementers as new or enhanced demand response programs, markets, and products are developed at the local, state, and regional level, as discussed in greater detail later in this section.
2. Develop a consistent message framework based on foundational market research conducted and compiled through the National Action Plan that utilities, load- serving entities, or other demand response

providers and stakeholders can independently tailor to their own local audiences to foster greater demand response participation and to supplement their existing communications with customers. Messages could include information on the versatility of demand response resources for a variety of applications beyond lowering peak demand, such as automating demand response at all hours of the day, using demand response to provide operating reserves and other ancillary services, and partnering demand response with wind generation and other variable renewable resources.

3. Give local demand response implementers the information they need to disseminate communications locally to residential and small commercial customers in areas that are implementing new or improved demand response programs, products, technologies, and incentives.
4. Leverage the success of the demonstration programs to spread information and successful program and product models that can be replicated in areas that plan to implement demand response in the near term.
5. Eventually disseminate communications to all localities, states, and regions as opportunities for demand response become more widespread.

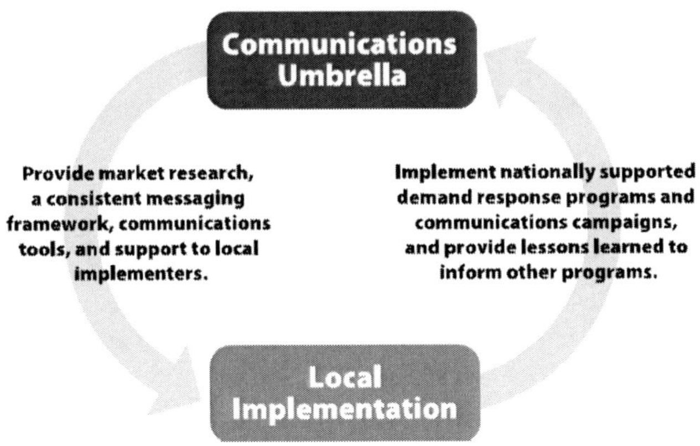

Figure 4. Interaction between the Communications Umbrella and Local Implementation.

The Communications Umbrella would focus on three target groups: 1) states, policymakers, and national partners who will make decisions about demand response implementation; 2) local demonstration program sites that have implemented or are implementing demand response and wish to have hands-on assistance in their outreach to customers; and 3) other demand response programs, including programs that are implementing demand response and wish to be fully autonomous in their communications with customers. The program would be most effective if it first helps stakeholders build decision-maker support and technical capacity for demand response implementation, then educates residential and small commercial customers through integrated communications in the localities and states that adopt demand response programs or where regional markets for demand response exist. Education would be conducted both in partnership with local demonstration program sites and by providing support materials to any sites that are not demonstration sites but want to implement demand response or supplement existing communications.

The responsibilities of the Communications Umbrella fall into two subsets: program development and message framework development. Communications Umbrella program development would involve compiling case studies and successful outreach tactics, identifying demonstration program sites, and conducting program evaluation. Message framework development involves conducting both a scan of existing research and new, foundational market research; developing adaptable customer-friendly messages; positioning demand response as a component of smart energy use; developing communications materials and toolkits; and developing support materials tailored to large commercial and industrial customers. As explained previously in Figure 4, some of these tasks will occur concurrently.

Program Development

The Communications Umbrella would first study how local entities have communicated about demand response implementation across the country and internationally. This step might include messages, focus group feedback, advertisements, and other components. From there, the Coalition would oversee the selection of demonstration program sites from among those states, local load-serving entities, utilities, and other demand response providers that have requested support and — once their work begins — evaluate their success.

Compile case studies

Though several entities have already compiled some data on demand response program implementation,[55] there is no single repository for customer research, message testing, behavioral data, and the like. The Coalition would oversee the collection of such nonproprietary data from as many sources as possible, given resource availability, and would make the data available through the Web-based clearinghouse, as discussed in Section 2.3. The best practices compiled through the clearinghouse would aid local programs in their own implementation efforts and, more importantly, inform the foundational customer research.

Identify demonstration program sites

Identifying demonstration program sites or markets is essential to the structure of the Communications Umbrella development process whereby the results, experiences, and successes in one location provide input and feedback to inform others. The Coalition would oversee the development of a method for choosing sites from a pool of applicants; it would consider factors such as the willingness of local stakeholders to be vocal demand response leaders, the receptivity of local customers to participate in programs, and the availability of local resources that can be used to fund communications related to demand response. Demonstration programs will be discussed more fully in the "Local Implementation" section.

Conduct program evaluation

The Coalition would oversee the development of metrics to evaluate the impact of particular outreach tactics and adjust the best practices and recommendations moving forward.

Message Framework Development

The Communications Umbrella would develop the broad message framework and outreach strategies for the national communications program, which could then be customized by local stakeholders. For example, as discussed below, the Communications Umbrella could provide a template one-page message document outlining to customers what demand response is and how it impacts them, and stakeholders would have the ability to edit the document, change the wording as necessary for their circumstances, and add their own program logos to it, as well as reformat it entirely or use only part of it in their other communications with customers.

Conduct foundational market research

Foundational market research would be used to develop a nationally consistent message framework for demand response, similar to the ENERGY

STAR® program.[56] The national message framework would facilitate customers' understanding of demand response and its possible benefits. Regardless of the eventual funding level available for implementation of the National Action Plan, foundational market research should be considered a priority item, because it would inform the national communications program of the most effective demand response messages and outreach tactics. Additionally, numerous stakeholders commented that conducting foundational market research should be a high priority for the national communications program.

As a starting point, research would look at the success or failure of existing demand response programs and identify lessons learned. It would also involve original research with residential and commercial electricity customers to identify which incentives and messages each customer class would find most persuasive to modify their energy usage. Much of this research would be relevant nationwide. However, research would also help to identify how best to tailor the messages to account for regional differences in the structure of electricity markets, critical peak periods, demand response program design, pricing models, and structure of utilities.

Formative research would help in assessing customer assumptions and understanding of key issues, conscious behavior, and what motivates customers to change their behavior. This foundational market research could include quantitative surveys of residential, commercial, and industrial customers, as well as qualitative research such as focus groups in numerous locations across the country and among varying customer classes. Key research questions that could be explored include:

- How might messages be tailored to reach different classes of potential demand response customers with different demand response capabilities?
- What messages are most effective to help customers understand what demand response is?
- What messages are most effective to help customers understand demand response's benefits?
- What level of understanding of demand response do customers need in order to participate? Do they need to understand the concept fully, or is it enough to give them easy tools and instructions and inform them of demand response's potential benefits?
- Are financial incentives such as electric bill savings or rebates the most powerful motivators?

- Are customers more inclined to change their behavior if they learn that reducing peak demand helps the environment? For example, is there a difference in response between paying a customer not to consume (but no penalty if they continue to consume) and building in a discount that the customer surrenders if they do not respond (peak time rebate vs. critical peak pricing (CPP))?
- How much inconvenience are customers willing to endure if their provider asks them to be part of a demand response program or product?
- Who should broadcast the demand response event for it to have maximum impact?
- Do messages that promote energy efficiency or energy conservation conflict with or support demand response messages?
- Do customers react differently to messages that talk about demand response as preventing crises versus messages that frame it as a purely positive step as we increase the use of renewable energy sources?

Customers' answers would help the national communications program shape its messages in all outreach support materials and ensure that the right messengers deliver those messages. In addition, having the answers to these and other research questions would benefit entities interested in investing in demand response programs, products, technologies, incentives, and infrastructure.

Foundational customer research would also provide insight on the effectiveness of linking the concept of demand response to other smart energy use concepts (e.g., state renewable portfolio standards, climate change, energy efficiency, regional markets, smart grid), compared with the effectiveness of educating customers about demand response as a wholly separate concept in order to achieve peak use reductions.

Finally, the research would help to determine the best market segments to target to achieve the greatest behavioral shifts for the lowest cost. Because virtually every person and every business uses electricity, the potential audience for this program is the entire adult population, which would be prohibitively expensive for any program to reach. Therefore, "zeroing in" on smaller groups would help to ensure efficient message delivery and avoid wasting time and resources on audiences who are unlikely to change their behavior.[57] This knowledge would inform the overall strategic and tactical recommendations developed through the Communications Umbrella for use by local programs.

Develop adaptable, customer-friendly messages

The majority of stakeholder comments stated that "demand response" is not a customer- friendly term. Another objective of the Communications Umbrella would be to reframe demand response with new, more customer-friendly terminology that would better explain the concepts of energy use reduction, load shifting, and managing energy usage in response to price or non-price signals, in conjunction with other smart energy use principles. The Coalition would oversee research to help determine the terminology that best conveys to customers what demand response is and how it can benefit them.

In addition to reframing the demand response concept, the Communications Umbrella would develop a message framework with persuasive, adaptable messages aimed at various audience segments, all of which could be tailored by interested local stakeholders. The program would segment the audience into discrete groups based on their varying motivations and circumstances, and messages would differ slightly for each segment. Messages must first define the problem that demand response programs could help to resolve; messages could then establish the urgent need for demand response implementation and explain appropriate ways to modify behavior. How best to communicate the problem and its solution may vary by audience. To simplify this notion, Table 1 provides examples of message categories for diverse audiences. The Coalition would oversee the development of messages for various audience segments within the message framework, with research providing insight into the most persuasive ways to convey the content listed in the Table.[58]

Table 1. Segmentation of Message Framework

Audience/"Who"	Message Content/"What"	Messenger	Delivery/"When"
Key stakeholders	Be a part of the Coalition that will lead the demand response program	Federal/state/ local governments, utilities/ load- serving entities/ other demand response providers, RTOs/ISOs/ commercial and industrial customers,	Immediately/ continually (as outreach begins in new states)
		consumer advocates, existing coalitions of these organizations	
All stakeholder leadership	Implement your demand response goals with our support	Coalition	Short-term/ongoing

Audience/"Who"	Message Content/"What"	Messenger	Delivery/"When"
Stakeholder leadership in states, localities, and regions with demand response	We can support and supplement your efforts with technical and communications assistance	Coalition	Short-term/ongoing (continually as new states adopt demand response)
Stakeholder leadership in states, localities, and regions receptive to demand response	Implement demand response with our support; once implemented, we can support and supplement your efforts	Coalition	Short-term, medium-term, long-term
Stakeholder leadership in non-demand response states	Look at success in other states and localities as a potential roadmap; we can help you achieve your unique goals	Coalition	Medium-term, long- term (continually as success stories and best practices emerge)
State and local governing officials	Support customer interest in implementing and receiving benefits of demand response	Coalition	Short-term/ongoing
Utilities, load-serving entities, and other demand response providers	Use our support to communicate with customers and provide additional sources of ancillary services and tools to reliably integrate variable resources	Coalition	Short-term/ongoing
Large commercial and industrial customers	Implement demand response because it can provide widespread benefits, save you money, and be directly beneficial to you as a participant in wholesale markets	Coalition, utilities/load-serving entities, public utility commissions, other demand response providers	Short-term/ongoing
Current residential and commercial customers with access to demand response programs	Join the program because it can provide widespread benefits (e.g., you could earn a cash rebate, protect our power supply, and promote the use of clean energy)	Utilities/load- serving entities, public utility commissions, Coalition, other demand response providers	Medium-term, long- term (continual roll-out as new states and localities adopt demand response)

Table 1. (Continued)

Audience/"Who"	Message Content/"What"	Messenger	Delivery/"When"
Low-income, elderly, and disabled customers with access to demand response programs	Join the program because it will provide you with benefits (it could save you money and could protect your power supply) and will not raise your costs or hurt your quality of life	Utilities/load- serving entities, public utility commissions, Coalition, other demand response providers	Medium-term, long- term (continual roll-out as new states and localities adopt demand response)
Future customers	Demand response is an important way to reduce energy use during peak periods, protect the environment, and promote the use of clean energy	Coalition, utilities, load-serving entities, schools, other demand response providers	Medium-term, long-term
Customer advocates	Support demand response nationwide because of its widespread benefits (e.g., less need for future power plants, safer power supply, potential money savings)	Coalition, other demand response providers	Short-term/ongoing
Renewable resource developers	Partner with demand response for ancillary services	Coalition	Short-term, medium-term, long-term

The Communications Umbrella would develop comprehensible, research-based messages aimed at the audience segments outlined above. It would also craft messages that would increase awareness, encourage customers to modify how they consume electricity, and assist customers that want to participate in available demand response programs.

For example, when addressing residential customers in areas with available demand response programs, products, or markets, a generic message framework might inform them about the following topics:

- **Variability of electric prices and sources.** Inform customers that the cost of generating electricity varies by time. During "peak periods" when there is high electricity demand, electricity is more expensive to produce. Also inform customers about which generation sources are used by time of day and during peak periods.

- **Urgency of demand response.** Communicate the urgency of the need for demand response, load shifting, energy efficiency, energy conservation, etc.
- **Actions to take.** Educate customers about specific actions they could take at critical peak periods to reduce electricity use, and publicize the potential benefits of participating in demand response. These messages would highlight the importance and value of taking action by tying that action to the most effective customer motivations, such as potential electricity bill savings or environmental benefits.
- **Minimizing negative impacts.** Provide reassurance to customers that any negative consequences of demand response implementation have been anticipated and would not pose a threat to them. These messages would lessen customers' concerns and prevent customer backlash, in concert with messages about the need for demand response, the specific actions to take, and the benefits of taking them.

When addressing large commercial and industrial customers, the message framework might inform them about:

- **Compensation.** Inform customers that their actions can not only reduce bills, but, in some parts of the country, also result in compensation opportunities.
- **Opportunities.** Educate large commercial customers about the role they may be able to play by participating in local utility programs or in regional markets (such as participating in demand response programs that could help to integrate variable energy resources like wind generation into the system), if applicable.

Foundational research, message testing, and Coalition input will help to determine whether these messages or others are appropriate and whether they motivate customers to change their behavior.

The Communications Umbrella also could work with a sampling of local demand response implementers to develop messages to educate customers about specific actions they could take at critical peak periods to reduce electricity use.

The intended outcome of the message development and testing process is to create a message framework that local stakeholders can adopt to communicate with customers, or that they can use to supplement or modify their own communications.

Following the finalization of the message framework, the Communications Umbrella would develop template advertising materials — including a template TV advertisement, a template radio advertisement, a template online advertisement, a template print advertisement, and template direct mail — that local stakeholders could use in their outreach efforts. To save on costs, the materials could be designed so that they would be produced once but still be customizable with graphics and website URLs for a specific state, load-serving entity, utility, or other demand response provider's demand response program.

Position Demand Response as Part of Overall Smart Energy Use

A core question in the development of messages for the Program is how best to position demand response in relation to other smart energy use concepts, such as renewable energy, load shifting, energy conservation, energy efficiency, and smart grid. Which relationship best aids customer understanding? In addition, messages should consider providing linkages to known, evolving, and emerging technologies (e.g., automated demand response such as Auto- DR for commercial and industrial facilities) and opportunities in retail and wholesale energy markets. Anecdotal market research to date shows that consumers already confuse these concepts of energy conservation and energy efficiency, so positioning demand response as a new twist on existing concepts may create further confusion. Though foundational customer research would provide the best insight, it may be necessary to link demand response with other smart energy use principles to increase customer understanding, or even to use other smart energy use terms to describe demand response.

Linking the *concept* of demand response with other smart energy practices would provide a strategic framework to integrate the education and promotion of demand response with other federal government or state-based initiatives to educate customers on smart energy use or sustainable energy conservation practices. For example, under the ENERGY STAR® program, DOE and EPA jointly market cost-effective opportunities for customers to receive the same level of service by using fewer kilowatt-hours of electricity, saving money and reducing carbon emissions.

The Coalition could work with existing renewable energy, energy efficiency, smart grid, and other related initiatives to oversee the development a comprehensive message. Doing so could persuade customers to take action and alter their energy use patterns, while meeting the communications objectives of each of the individual programs.

Develop communications materials, tactics, and toolkits

To support all stakeholders implementing demand response initiatives, the Communications Umbrella would develop a suite of toolkit materials that stakeholders could use to create their own customized communications programs or to supplement local communications. Toolkit materials (explained in greater detail in the "Support for Other Demand Response Programs" section) would include message research data and analysis, guidance on demand response messages, strategies and tactics for effective audience targeting and outreach, and customizable advertisements and other communications materials.

Develop support materials tailored to large commercial and industrial customers

Residential and small commercial customers are a very important part of the audience for demand response, and reaching this audience segment effectively would have the greatest positive impact on broad public opinion surrounding demand response. But effective communications to residential and small commercial customers cannot begin until demand response programs, products, technologies, and incentives are in place in their states or localities.

Unlike residential customers, large commercial and industrial customers generally have greater awareness and knowledge about energy and actions that can be taken to adjust their electricity consumption. In addition, both retail and wholesale demand response programs are already available in many places throughout the country in which these customers may participate. Because of this greater sophistication on energy issues, existing program availability, and the likely cost-effectiveness of these programs, outreach efforts to these customers could result in immediate benefits associated with demand response participation by this sector.

Working at the national and local levels, depending on the customer, the Coalition would coordinate outreach to large companies to inform them about demand response and available opportunities to provide it. This outreach would showcase how demand response can be made available through distinct messages and template materials geared toward large-scale customers.

The Coalition would work with state and local stakeholders to provide materials about best practices and model program guides that would help large customers participate in demand response programs offered by a utility, load-serving entity, other demand response providers, or RTO/ISO markets. Template slide show presentations could be used to inform customers and enlist their support. A template press release could be customized to congratulate and thank large commercial and industrial customers for investing in demand response.

Local Implementation

Support for Demonstration Programs

Purpose	Advise states and localities on communications best practices as they adapt the Communications Umbrella to increase exposure of and participation in demand response in their own localities. Provide logistical support and coordination for communications around local demand response implementation.
Expected Outcome	Tailored communications that integrate the broad, national messages with the local program's needs and unique characteristics.

A core element of the national communications program is the implementation of a broad strategic and message framework through a local demonstration program approach.

This local implementation demonstration program approach would first roll out in areas that have requested the national communications program's support and have already implemented or are on the verge of implementing pilot or full-scale demand response programs and products. The demonstration program approach would then spread (as requested) to areas that plan to implement programs and products in the near term, and would eventually become national in scope as implementation becomes more widespread. The national communications program could coordinate with DOE's Recovery Act stimulus grants. For instance, several of the recipients of the stimulus grants will be conducting controlled experiments of the response to the smart metering and dynamic pricing. These experiments could serve as examples within the demonstration program approach.

Using broad, nationally developed messages, the Communications Umbrella would advise local load-serving entities, utilities, and other demand response providers that have agreed to participate and have been selected as demonstration programs on how to translate the Communications Umbrella messages and strategies to their own local demand response initiatives, seasonal peak load patterns, and any other regional factors.

The strategy for communicating messages around local demand response implementation is an integrated communications program, not reliant on one particular communication method or medium. The integrated marketing elements supported by the national communications program for a sample demonstration program could include:

- **Public relations.** An important tactical component of a local communications program would be a comprehensive news media outreach and public relations strategy. This strategy would build public and policymaker support for demand response in conjunction with direct communications outreach to states and policymakers (discussed later in the Direct Outreach to States, Policymakers, and Partners subsection) and the technical assistance to states discussed in Section 2.1. It would also educate customers by garnering news coverage about demand response. By communicating with members of the news media about demand response's benefits, the need for it, and the ease of implementation for the customer, positive news stories and editorials would help to shape public opinion in favor of demand response. In addition, appearances by experts on talk shows would give more customers a better understanding of demand response.
- **Advertising.** If local resources allow, advertising or public service announcements (PSAs) could be a key part of local communications efforts, with the national communications program providing the template advertisements or PSAs that could be tailored to reference local programs and information. Media buys for paid advertising could be funded by states or localities, utilities, load-serving entities, other demand response providers, or a combination of these groups. If paid advertising funding is unavailable, the PSA version can be used to attain free placements of the same messages. To reach the broadest range of customers in each market, depending on which customer segments the foundational research indicates offer the best opportunity, advertising might run on:
  - Television
  - Radio
  - Outdoor media, such as billboards and advertisements on buses
  - Direct mail
  - Internet

- Emerging media technologies, such as cell phone text messaging

The Communications Umbrella would also develop attractive, customer-focused educational support materials, which could be tailored with local information and would be available digitally for stakeholders' use (e.g., brochures, videos, PSAs, a reference website).

- **Viral[59] and grassroots marketing.** In addition to a local mass media program, a fully integrated communications program could also include a grassroots approach that leverages emerging social networks in the digital space. A viral and grassroots marketing outreach might include:
  - The creation of groups within social media sites like Facebook and Twitter, as well as new sites as they gain popularity.
  - The engagement of third-party "demand response champions" who can blog or tweet about it.
  - Development and posting of viral videos that demonstrate the potential benefits of demand response or, alternatively, what the future might be like without demand response.

  This type of approach could potentially tie into a national communications program over time, as the social network around the groups, champions, and videos grows larger and more geographically diverse.

Example: Grassroots Marketing

In 2007, UNICEF created a grassroots initiative called the Tap Project, which raises money during World Water Week with the goal of providing clean water to children around the world. The Tap Project began in 2007 as a demonstration program in New York City and has now expanded to all 50 states. The national campaign develops the broad campaign message platform, an annual theme, and template materials, and then hands them over to individual communications teams in each of the targeted cities. Each city's team translates the messages and theme into its own campaign and launches that campaign through PSAs, news media outreach, local events, and extensive online viral and grassroots marketing. The result is a series of localized communications efforts with their own unique characteristics.

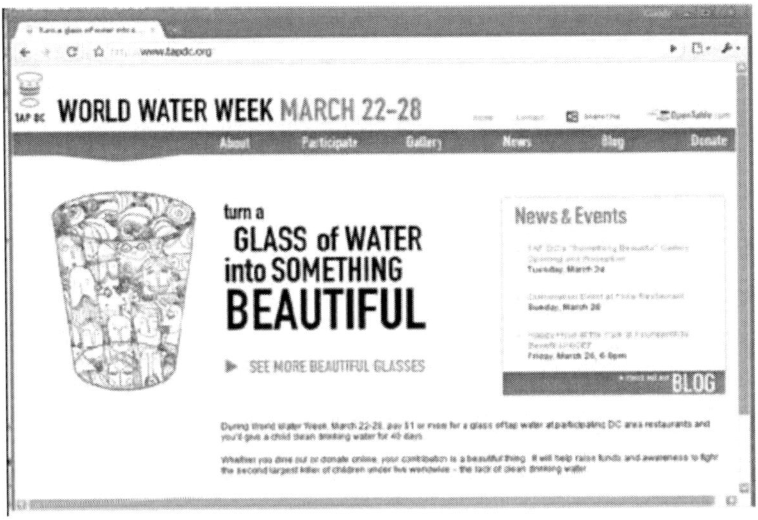

Support for Other Demand Response Programs

Purpose	Provide tools adapted from the Communications Umbrella so regions, states, and localities can employ the messages and tools most useful to them in informing customers and stakeholders about demand response.
Expected Outcome	A broad spectrum of toolkit materials that can supplement an entity's existing communications efforts, add an additional program layer, or serve as a new program where previously there was none. Guidance on strategies for effective customer outreach.

To support the efforts of utilities, load-serving entities, and other demand response providers that do not directly participate as demonstration programs, the Coalition could oversee the development of template toolkit materials, as mentioned earlier in this section, enabling those entities to manage their own local demand response communications programs. The materials in the toolkit would be customizable to account for a local program's name, logo, website URL, and needs. The toolkit materials would be designed to be "plug and play" so that local entities would be able to use all available messages and materials or choose which elements to use. Local stakeholders, states, or localities could fund local implementation of toolkit-based communications efforts.

Toolkit materials could include:

- A foundational customer research primer covering customer behavior, responses to various terms and messages around demand response, findings on how best to position demand response as a part of smart energy use, and analysis of how best to segment audiences to achieve maximum results.
- Sample messages, which can be tailored with local program information and segmented by audience type, seasonal peak load pattern, general demand response program and pricing design, and other variables.
- Template talking points for local utilities, load-serving entities, and other demand response providers, segmented by the variables outlined above.
- Template PSAs that can be run on local television without charge, as well as guidance on how to get PSAs placed on local broadcast and cable networks.
- Template press materials, including press releases introducing a new program or announcing the introduction of new technology or pricing, frequently asked questions for journalists about demand response, and how-to guides for media outreach.
- Print-ready materials such as brochures and flyers, designed to be printed with local logos, program names, and website URLs inserted.
- Advertising for TV, print, radio, outdoor, and online media, all of which can be tagged with local program information.
- A sample integrated communications program plan, outlining how best to maximize the local program's impact with a combination of various outreach methods, as well as guidance on how to select media to reach the most receptive audience segments.

Example: Communications Toolkit

The Covering Kids & Families campaign provided free toolkits and message guides to anyone interested in reaching out to families about children's health coverage. The campaign developed tools such as the comprehensive "Back-to-School Campaign Action Kit," which provided templates and tools to do everything from writing and sending a press release to conducting a local enrollment event and press conference. Another toolkit provided step-by-step instructions on customizing and placing a PSA. These toolkits even enabled organizations that were not official grantees or demonstration programs to be a part of the larger campaign and get the message out in the most convenient way.

Direct Outreach to States, Policymakers, and Partners

Purpose	Provide a general framework of support for demand response implementation programs taking place across the country.
Expected Outcome	Broader education for policymakers and stakeholders necessary to the successful implementation of demand response, and national and regional partnerships supporting local efforts.

Although the bulk of the national communications program activities — particularly public education — would take place at the regional and local levels, there is an important role for certain national activities that would amplify the impact of local communications.

Communications Assistance to States and Policymakers

Support to states, localities, and regions would be another key part of the national communications program. Communications programs or meetings would be provided, possibly as part of regional workshops, to educate state and local governing officials on how to communicate with customers about demand response. Such educational programs would offer an opportunity to share findings from consumer research and demand response experts, provide information on messages to engage various stakeholder and consumer audiences, and provide tools to ensure state and local governing officials can begin reaching out to customers with educational information about demand response programs, products, technologies, and incentives. Because support from state and local governing officials is important for demand response implementation, efforts to implement demand response programs and products locally may benefit from the Coalition coordinating the provision of communications assistance to state and local governing officials and stakeholders to allow them to properly assess and understand demand response in an unbiased manner. Such communications assistance might include: 1) communications education and message guides that outline the need for demand response action, 2) national spokespeople to inform media and state and local governing officials, and 3) assistance in building support for demand response among the general public. This assistance would enable policymakers and others to inform customer groups about demand response opportunities and capabilities they can incorporate into their businesses, industries, and homes, ultimately boosting customer acceptance of demand response.

Partnership Development

The national communications program would foster development of national and local partnerships between the Coalition and 1) corporate partners whose customers are making decisions about electricity usage; 2) organizations representing various customer classes, such as commercial and industrial customers, certain subsets of residents, etc.; and 3) trade associations or other groups that would benefit from partnerships, e.g., the pairing of variable resource development with dispatchable demand response.

Both manufacturers and retail sellers of large home appliances that could be demand response– or smart grid–enabled would be ideal partners for the program. Because these corporations regularly interact with the public (e.g., when customers consider purchasing new appliances), achieving some program and message visibility at the point of purchase would be extremely advantageous and would reach the audience when it is most receptive. Similarly, partnerships among trade associations, such as those representing industrial customers and those representing wind resources, might help states meet renewable portfolio standards.

Partnerships between for-profit corporations and nonprofit cause-related programs are common. Corporations are eager to participate in such partnerships if they believe their participation could have positive business repercussions or improve their image through association with a worthy cause. For example, a home improvement retailer chain may be interested in partnering with the national demand response program to 1) associate the company with an energy-saving initiative that shows its customers that it is forward-looking and concerned with environmental health, and 2) increase sales of demand response– or smart grid–enabled appliances. Partnering with such a company could make demand response issues more visible to customers and create additional touchpoints between the customer and the issue beyond traditional media outlets.

In terms of organizational partnerships, groups that represent businesses, industrial customers, and subsets of the public at large would be excellent program partners, as they provide access to millions of key business and household decision-makers through their communications networks. Such groups include trade associations and consumer advocacy organizations. These types of partnerships would allow the Communications Umbrella to target the groups or geographic regions that can make the best use of demand response to reliably and economically meet goals and objectives at the local, state, regional, national, or individual levels. Coordination with key associations representing these customer segments could increase the effectiveness of a communications strategy while reducing costs.

Corporate and organizational partners would be selected on the basis of their reach within key audience segments, their relevance to advancing the issue, and their ability to commit resources to support the effort. They also could support message outreach to groups that might not otherwise learn about the benefits of demand response programs, or that might not have the information necessary to improve their lives by participating in demand response programs.

Example: National and Local Communications Partnership

The Covering Kids & Families campaign sought to enroll eligible children and families in public health insurance programs. To get these children enrolled, the campaign had to reach their parents. Through the development of corporate partnerships, the campaign gained millions of dollars of in-kind placement in retail environments where parents are most likely to be thinking about their children. In a partnership with Giant Foods on the East Coast, the campaign secured placement of an enrollment PSA on the sides of milk cartons during the back-to-school time period. In a partnership with CVS/Pharmacy, the campaign ran PSAs on the prescription information forms for anyone under the age of 18. These partnerships involved no direct costs for the campaign but provided enormous added visibility for its messages.

2.3. Tools and Materials For Use by Customers, States, and Demand Response Providers

Section 529 of EISA requires that the National Action Plan develop or identify analytical tools, information, model regulatory provisions, model contracts, and other support materials for use by customers, states, utilities, and demand response providers. Therefore, in this section, the National Action Plan identifies tools and materials that can be used by customers, states, and demand response providers. The Coalition or the federal agency would oversee the development and management of these tools and materials. The level of effort put into these tools and materials will depend on the funding and other resources available for implementation. With limited resources, the elements proposed in this section would largely be enhancements of existing tools and support of ongoing efforts. New and innovative tools and materials would be developed if more resources can be applied.

The development and enhancement of tools and materials will serve several objectives. For instance, states not currently engaged in demand response will have access to a comprehensive source of information that could help them decide what types of demand response programs are appropriate. States that are already engaged in demand response will be able to share their experiences and gather more information about how to measure results and improve their programs. Utilities will be able to examine other pilot program design and results, while consumers will have tools to help them decide if they would benefit from demand response programs. All of this information will help optimize program benefits.

Based on stakeholder input and further research, the National Action Plan contains two general categories of action for the development of tools and materials:

1. Develop a Web-based clearinghouse of demand response materials.
2. Develop or enhance demand response estimation tools and methods.

Develop a Web-based Clearinghouse of Demand Response Materials

Purpose	Provide easily accessible, accurate information on demand response for stakeholders.
Expected Outcome	Increased awareness of existing demand response experiences and their reported benefits and lessons learned, resulting in additional use of well- designed demand response programs.

There is strong support for developing an easily accessible website database, or "Web-based clearinghouse," to serve as the centralized location for collecting all publicly available information on demand response, including regulatory documents, program reports, general demand response information, and communications material. The goal for the Web-based clearinghouse will be to provide stakeholders and consumers with the most current information and analyses on demand response.

The Web-based clearinghouse may be included as part of an existing federal agency clearinghouse or may be privately funded and maintained by the Coalition. In either event, the Web-based clearinghouse for demand response would include information that stakeholders can use to make their own evaluations and judgments on the reports and information that the Web-based clearinghouse contains. For example, in addition to posting a report on a demand response pilot program, it would present guidelines, criteria, and metrics that can be considered when evaluating pilots. Additionally, the clearinghouse would be a useful tool for making technical papers (developed as a result of Coalition discussions) available to states and others.

The Web-based clearinghouse should build on and collaborate with existing efforts to collect data on demand response programs and reports. The most relevant ongoing effort in this area is the new DOE-sponsored Smart Grid Information Clearinghouse.[60] The Web-based clearinghouse for demand response could be coordinated with this effort, which was launched with a $1.3 million budget (provided from the Recovery Act) in September 2009.[61]

Based on preliminary information, the Smart Grid Information Clearinghouse may include a significant amount of the demand response materials intended for the Web-based clearinghouse in the National Action Plan.[62] The Smart Grid Information Clearinghouse is expected to launch publicly in September 2010.[63]

This Web-based clearinghouse can use other efforts as a model. One is DSIRE, the Database of State Incentives for Renewable Energy — another DOE-funded project for collecting and distributing information about renewable energy and energy efficiency efforts. The DSIRE website provides users with visual tools for understanding the current status of funding, standards, and state and federal incentives and mandates for renewables. It also contains information about financial incentives and regulations regarding energy efficiency, along with an extensive library of industry presentations and reports provided on a state-by-state basis. This information allows for easy comparisons of programs and policies across states and provides a simple means for understanding the "landscape" of energy efficiency and renewable energy policy in the United States. Stakeholders have identified DSIRE as a useful example of an online clearinghouse for information.

The California Measurement Advisory Council (CALMAC) maintains a website containing information on demand-side programs in California. The purpose of the CALMAC website is to provide "...a forum for the development, implementation, presentation, discussion, and review of regional and statewide market assessment and evaluation studies for California energy efficiency programs and demand response..."[64] The site's searchable database is helpful for finding relevant, up-to-date materials, particularly impact evaluations conducted by utilities and independent consulting firms. This type of information could be included in the Web-based clearinghouse.

The objectives of the Web-based clearinghouse would change over time. In the short term, the effort would be focused more on assembling the results of pilots and studies on the impacts and benefits of new dynamic pricing and smart grid-enabled programs. This information would help to establish confidence in the ability of these new programs to provide reliable benefits. In the long term, as programs become established through full-scale implementation, the focus would shift to identifying model programs and policies, as well as prioritizing findings from the research that has been conducted.

The following subsections describe the types of content that the Web-based clearinghouse could include.

Case Studies of State Laws, Regulations, and Tariffs

The Web-based clearinghouse would provide access to comprehensive case studies that include specifics about demand response programs. For example, information regarding state laws, regulations, tariffs, generation mix, forecasted load, technology employed, communications strategy, and retail cost recovery would provide helpful contextual information for understanding and comparing demand response programs. To the extent possible, actual laws, regulations, and tariff language would be provided.

State laws, regulations, and tariff language included in the Web-based clearinghouse could be used as a sample or template that stakeholders can adapt, based on their unique goals, preferences, and needs, in developing their programs and policies. Furthermore, any contextual information that is included would help stakeholders adapt the state laws, regulations, and tariff language to their unique circumstances.

For example, California's Energy Action Plan includes a model regulation for preferred loading order of resources: energy efficiency; demand response; renewable energy sources; and conventional generation options that could be included in the Web-based clearinghouse for other states to use as a model for similar legislation.[65] Also, California's Energy Action Plan set a target reduction in peak demand of 5 percent for 2007.[66] When the three investor-owned electric utilities only achieved 40 percent of that target in 2007, the utilities were asked to revamp their efforts. Simultaneously, CPUC approved business cases for the deployment of AMI by all three utilities[67] and then issued a landmark ruling making dynamic pricing the default tariff for all nonresidential customers who were part of AMI. Information about California's Energy Action Plan, coupled with updates about how it is being carried out, would help other states if they undertake similar efforts.

Connecticut provides another example of a regulation-related case study. The Connecticut state legislature passed legislation that required electric distribution companies to conduct integrated resource planning beginning on January 1, 2008.[68] The identified resource needs must first be met through all available energy efficiency and demand reduction resources that are cost-effective, reliable, and feasible.[69] Among other things, the electric distribution companies' procurement plans must specify the amount of demand-side measures, such as demand response, that can meet the capacity planning needs.[70]

Database of Pilots, Programs, and Markets

There are different types of demand response programs across the country. The Web-based clearinghouse should include a database of both pilot and fully implemented demand response programs. The database should include general descriptions of each of these programs and, to the extent possible, reports that summarize each program's results, lessons learned, and benefits and drawbacks.

Information on the programs should be supplemented with marketing materials, surveys of customer satisfaction, and the names of contractors or vendors that provided enabling technology and devices that supported demand response. These materials would provide useful qualitative information that stakeholders could use to enhance their own programs. Furthermore, the lessons learned from this information could supplement the communications materials that are developed as part of the communications program (see Section 2.2).[71]

The existence of a central collection of reported impacts from existing demand response programs and products would help shed light on the ability of demand response to provide consistent, measurable impacts and would also serve as a resource for benchmarking newly reported impacts from demand response program offerings.

The Web-based clearinghouse should include a broad mix of programs to serve as case studies in program design, implementation, and evaluation. Examples include new pilot programs run by utilities such as Consumers Energy in Michigan, Commonwealth Edison in Illinois, Baltimore Gas & Electric in Maryland, and Northeast Utilities in Connecticut, as well as mature programs such as Detroit Edison's Direct Load Control Interruptible Air Conditioning[72] program and California's "Flex Your Power" campaign to promote customer awareness of pending energy shortages and encourage reduced energy use.[73] Users of the Web-based clearinghouse would be able to find information on each of these programs and, if they exist, reports evaluating the impacts of each program. With this information, users would be able to compare and contrast the value of each program, which would help to inform their decisions on how to implement their own programs.

Quantitative and Qualitative Summaries

Detailed supply-side information is contained on the U.S. Energy Information Administration's website. The Web-based clearinghouse will aspire to provide a similar level of granularity in demand-side information. Examples of data that could be collected include a listing of each demand response program and product being offered, peak demand impacts from demand response, costs of the programs, and participation rates. One useful way to organize these data would be by utility, load-serving entity, state, and RTO/ISO.

Recently, FERC has been collecting some of this information through its survey of demand response programs. The results of this publicly available survey provide the basis of FERC's Assessment of Demand Response and Advanced Metering reports.[74] Additionally, the North American Electric Reliability Corporation (NERC) collects related data at the wholesale level and recently initiated an effort to improve its demand response data collection methodology.[75] The results of these efforts would be included as part of the Web-based clearinghouse.

In addition to quantitative data, the clearinghouse could also collect detailed qualitative information. For example, the clearinghouse could provide a profile of the regulatory environment in which the programs are being offered, showing whether the demand response program is implemented in a state that has rate decoupling and whether the demand response program is being driven by the utility, regulatory commission, RTO/ISO, other demand response provider, or other party. Another example is information on the variations in RTO and ISO demand response, such as what requirements the provider must meet in order to bid and the incentives to participate. Other helpful qualitative information could include the amounts and types of demand response being incorporated into transmission planning and resource planning, a summary of marketing and customer recruitment efforts, and key lessons learned.

The clearinghouse could also include or provide linkages to quantitative and qualitative information on demand response developed by other entities. For example, EPA could share information to help states and other stakeholders understand the potential effects demand response may have on emissions.

Measuring Program Results with Criteria and Metrics

The wide variety of demand response programs across the country leads to wide variation in the reports on the results and lessons learned. The Web-based clearinghouse should include criteria and metrics that help stakeholders understand and evaluate these reports, make informed judgments about the results the reports present, and make meaningful comparisons across the reports. For example, principles of pilot program design — internal and external validity, control and treatment groups, selection of participants, measurement of load and other data, and payment or other incentives—would be valuable to clearinghouse users.

Information for Program Implementation

The Web-based clearinghouse should provide information for stakeholders interested in implementing their own demand response programs. Unlike case studies and pilot results, which consider historical events and remain static, the information for program implementation could be updated with the results of case studies, pilot results, and utility smart grid and demand response business cases to highlight current and future issues in program design and implementation. Information would be provided on how to implement both dynamic rate design and non-rate-based demand response programs. The clearinghouse also should include information on available smart grid technologies, which will play a crucial role in demand response implementation.

A stakeholder must consider many options if planning to implement demand response using a dynamic rate.[76] The Web-based clearinghouse would provide information on options for both the dynamic rate selection and its specific design. For example, the strength of the price signal to customers (i.e., the differential between the peak and off-peak prices) indicates to the customer the magnitude of the incentive to reduce consumption when the price is high and relates to the customer's opportunity for significant bill savings. Program enrollment and peak reductions likely are dependent on the opportunity for bill savings. Finally, the complexity of the rate design will play an important role in a customer's ability and willingness to respond to a dynamic rate.

The clearinghouse would provide non-rate-based demand response information as well. For example, when designing direct load control programs, it is important to consider how signals will be sent to the load control devices. Furthermore, the devices can be both one-way (i.e., only receiving the signal to be controlled) or two-way (i.e., the device receives the signal to be controlled and sends a signal back to the dispatcher confirming receipt). The duration and frequency with which the controlled devices cease to operate are another example of useful information in this context.

The foundation of much of this type of information is publicly available through filings with state commissions. However, there is no central location for easily locating the most up-to- date materials on demand response program implementation and design stemming from prior efforts. Therefore, a coordinated effort to collect and organize these data, beginning with a comprehensive assessment of the data that are already available, would facilitate discussion of the issues facing stakeholders who are currently designing and implementing their own demand response programs.

Online Message Board Capability

In addition to serving as a central location for demand response reports and resources, the Web-based clearinghouse also should serve as an online message board where qualifying state agencies (e.g., state regulatory commissions, state energy offices), consumer advocacy groups, and utilities may ask for specific virtual or onsite demand response assistance with regard to program design, implementation, and/or evaluation. In conjunction with the demand response assistance program discussed in Section 2.1, experts could review the message board posts and offer assistance. The nature of this assistance might involve referring a caller to a website, providing applicable reference materials, or perhaps providing onsite technical help (for a day, a week, or longer).

Information about the Development of Standards and Protocols

The development of standards and protocols is important for creating an interoperable electric grid. However, because other organizations have already taken the lead in developing standards and protocols, the Coalition could oversee the tracking of the standards development process. Up-to-date information on progress in the standards development process would be made available through the Web-based clearinghouse.

The National Institute of Standards and Technology (NIST) has established a process and plan that are designed to establish smart grid interoperability standards per its mandate from EISA.[77] NIST has allocated $5 million of its Recovery Act funding for this purpose.[78] Over the past year, NIST launched the initial stages of its broad effort, including creating Domain Expert Working Groups (in conjunction with DOE) to concentrate on focus areas and developing an Interoperability Knowledge Base to promote shared understanding and terminology.[79] In addition, the North American Energy Standards Board (NAESB) and NERC have already undertaken standard and protocol efforts for the smart grid. The Coalition should design the Web-based clearinghouse to allow interested persons to track these efforts to develop standards and protocols, perhaps through coordination with the NIST Smart Grid Interoperability Panel.

There are many examples of standards and protocols that could be developed to support demand response. These include communication protocols for consumer demand information, data privacy and cyber-security standards, standards for plug-in hybrid electric vehicles and fully electric vehicles, and measurement and verification standards.

Standard communication protocols would allow information on electricity demand and prices to flow unimpeded from consumers to a utility, demand response provider, or wholesale market operator. All transmitted information also should be readily available to consumers, because these signals (which traditionally came in the form of calling or paging facility managers) instruct customers to reduce their load in response to market prices or grid reliability concerns. Current technology, such as the automated demand response program, allows system operators and utilities to interact directly with commercial and industrial facility control systems that are pre-programmed to take action based on a demand response signal, without manual intervention.[80] Standardizing demand response signaling could make it easier for facilities to respond to calls for load reductions and management, and thus potentially increase the amount of customer participation in demand response programs and products. An example of such a standard could be the Open Automated Demand Response (OpenADR) specification, developed by the Demand Response Research Center. OpenADR provides a standardized demand response communications and signaling infrastructure using open, industry-approved data models.

Demand response activities will generate a significant amount of new data about consumers' actions and energy usage. Protocols would be developed for the appropriate protection of this information, including standards for the entities that should, or may, have access to this information and specific conditions for accessing this information. Furthermore, it will be crucial for all newly generated data to be transmitted over secure lines of communication between system operators and demand response providers. Establishing a secure and open standard for the transmittal of private consumer data in addition to sensitive information relating to the overall electrical grid would be a significant step forward for states, utilities, other load-serving entities, and customers looking to invest in demand response products or other enabling technologies.

Nationwide standards for charging stations for PHEVs and electric vehicles also should be adopted.[81] These charging stations should have appropriate communications, metering and electric flow control, and standardized plug interface. These actions would facilitate the use of PHEV and electric vehicles for their variable storage potential to provide ancillary services to the electric grid. Nationwide standards would also reduce barriers to interoperability, such as the need to develop PHEVs and electric vehicles that meet standards that vary state by state.

Finally, development of measurement and verification standards is critically needed at both the wholesale and retail levels. Methods for measuring and verifying demand response reductions currently vary significantly across the country, and measurement and verification standards will increase confidence in markets. Efforts to standardize measurement and verification are already underway. NAESB is currently leading an effort that has produced a document describing model business practices for the measurement and verification of demand response.[82] NAESB is also working with NERC to ensure that NERC's demand response data collection effort conforms to a consistent measurement and verification standard. NERC's Demand Response Data Task Force is charged with developing a systematic approach for collecting and disseminating data on demand response programs, products, and services administered by retail and wholesale entities throughout North America. The efforts of NAESB and NERC should provide valuable tools and standardization to ensure that demand response is accurately measured and verified.

Develop or Enhance Demand Response Estimation Tools and Methods

Purpose	Determine the potential value and benefits of demand response solutions. Assess whether demand response solutions deliver value and benefits once deployed.
Expected Outcome	A set of tools and methods provided to regulators, operators, utilities, load- serving entities, other demand response providers, and consumers for conducting simulations, analyses, and evaluations of demand response programs that will lead to more cost-effective programs and optimal participation by consumers.

Analytical tools and methods would be developed or enhanced to: 1) help in the expansion of existing demand response programs and creation of new programs by valuating each program's benefits and costs; 2) advance the use of demand response to support reliable and efficient operations of wholesale transmission, energy, capacity, and ancillary services markets; and 3) better enable end-use consumers to optimize savings, possibly through automation, while enrolled in a demand response program.

The Coalition or the federal agency could oversee assistance to organizations that are developing or enhancing tools and methods. These organizations would begin by identifying well-established tools and methods already being used to evaluate program impacts and cost- effectiveness. Next, organizations would consider how these tools could be enhanced or consider areas where the development of additional tools would be beneficial. To do this, organizations should engage in new research identifying beneficial characteristics of existing tools or meet through conference venues to foster information-sharing on current successful examples. Based on the outcomes of either of these options, projects to modify existing tools and methods or develop new ones would be identified. The Coalition could help identify options such as the degree of customization. For instance, it might be better that some tools be developed at a generic level and then customized at user expense, while other tools include pre-developed options that allow users to customize them without additional expense.

These tools and materials would be tailored to circumstances in wholesale and retail markets, as well as restructured and nonrestructured markets with vertically integrated utilities, so as to capture the differing sets of incentives and rules for pursuing demand response. For instance, tools for analyzing retail demand response programs must reflect that these programs are run through load-serving entities or other demand response providers, and that consumers are usually faced with a tariff or contractual obligation for participation in demand response programs. Conversely, tools for analyzing wholesale demand response programs must be built to reflect that participants are bidding resources into a market, possibly setting the RTO's locational price.

In addition to helping support full utilization of demand response resources in deployment, these tools and methods also could be structured to support energy efficiency programs. This could be achieved by making the tools and methods more flexible and not focused solely on demand response. In particular, the tools and methods could include the option to calculate the impacts, benefits, costs, etc., of utilizing more energy-efficient equipment. The integration of energy efficiency with demand response in the creation of tools and methods will advance the objectives of both types of demand side resources.

The following subsections suggest the types of tools and methods that could be developed to support demand response.

System Impact Estimation and Prediction Tools

System impact estimation and prediction tools are important because they allow stakeholders and policymakers to estimate the effects of demand response programs, products, technologies, and incentives on system resource requirements, wholesale prices, and emissions of carbon dioxide and other air pollutants before spending money on pilots and/or full implementation. For some types of demand response programs, baseline predictions on consumer demand would need to be calculated. Tools and methods for conducting these calculations should be developed.

Tools that can analyze some aspects of potential demand response impacts already exist. As part of the resource planning process, most utilities and ISOs/RTOs operate capacity expansion models through which demand response and energy efficiency measures can "compete" with supply-side resources to determine the optimal mix of resources for the system. In this case, the federal agency or the Coalition should facilitate a deeper understanding of how demand-side programs can be accurately integrated into what traditionally have been more supply-side-centric modeling frameworks.

Additionally, through the National Assessment, FERC staff developed a tool for estimating the state-level potential for demand response under a variety of scenarios.[83] This tool was the basis for the estimates of demand response potential provided in the National Assessment and has served as a valuable starting point for individual states and other entities to conduct more detailed and tailored assessments of their own.

To support the National Action Plan and bridge the gap between estimates of demand response potential and existing capacity expansion plans, FERC staff is in the process of developing a quantitative estimation tool that, given a capacity expansion plan, will yield information on plant dispatch and energy margins, reserve margins, and unused supply. For a brief description of the tool under development, see Sidebar 3 at the end of this section.

The InterAct tool, developed by Pacific Gas & Electric Company, is an example of an existing demand response tool that has been developed for customers. This free online tool is available to any large customer (200 kilowatts [kW] or greater) with 15-minute-interval load shape data and temperature readings to guide decisions related to demand response. InterAct's Energy and Cost Analysis module provides customers with tools to create graphs and charts, conduct statistical analyses, and export data. These options allow the consumer to closely track and understand their electric demand and usage. This tool helps consumers optimize their participation in demand response programs by notifying them of events via multiple media, allows consumers to easily accept or decline a curtailment notice, and gives consumers actual demand reductions and incentive payments after the event day.[84]

Another example comes from the Building Energy Software Tools Directory.[85] The Demand Response Quick Assessment Tool, developed by Lawrence Berkeley National Laboratory, helps commercial customers predict energy and demand savings for various demand response strategies. It allows users to run simulations based on a set of assumptions and variables.[86]

A final example of an analysis tool comes from the Rocky Mountain Institute (RMI). RMI developed a dispatch model of what it calls the Next Generation Utility.[87] The model is designed to provide insight on the potential of specific dispatchable demand response resources and technologies (e.g., for air conditioning, lighting, and plug loads) to reshape the load profile to better match renewable output. RMI's modeling focus has been on analyzing the combined effects of these resources in an electric system with high penetrations of wind and solar energy. The capabilities of the dispatch model with respect to demand response are: modeling traditional load-curtailment demand response; modeling thermal storage strategies such as ice storage and pre-cooling buildings to analyze consumer savings and electric system impacts; and analyzing driving and charging patterns for fleets of plug-in vehicles. RMI is also incorporating the hourly dispatch methodology into an integrated system planning tool for utilities and regulators to use for modeling the financial and operational characteristics of future scenarios. This expanded functionality will include modeling the impacts of investing in demand response along with renewables to minimize reliability concerns.

Example: InterAct Tool

The InterAct tool provides summary statistics regarding usage and load.

Summary Statistics

Energy Statistic	Value
Weekday Energy Usage (kWh)	283,533.36
Weekend Energy Usage (kWh)	117,903.12
Total Energy Usage (kWh)	401,436.48

Load Statistic	Value
Load Factor	65.16
Weekend Load Factor	69.02
Weekday Load Factor	71.34
Avg Demand (kW)	539.57
Min Demand (kW)	0.00
Max Demand (kW)	828.00
Power Factor At Maximum Demand	
Weekday Max Demand (kW)	828.00
Weekend Max Demand (kW)	647.04

The InterAct tool summarizes usage and demand data in graphical form with temperatures overlaid.

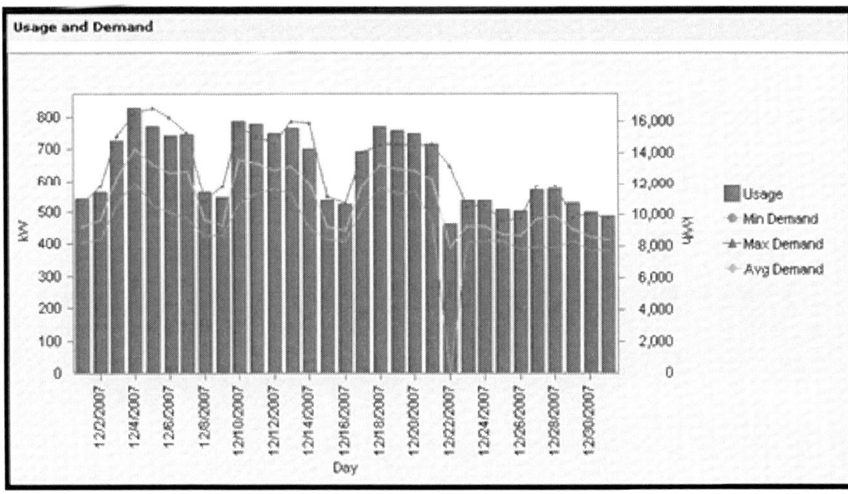

Example: Demand Response Quick Assessment Tool

The Demand Response Quick Assessment Tool has an interface that allows users to set the assumptions and variables associated with a building.

Example: RMI Dispatch Model

The RMI dispatch model illustrates how demand response can shift demand to times when renewable generators are producing power rather than dispatching expensive and inefficient quick-ramping generators.

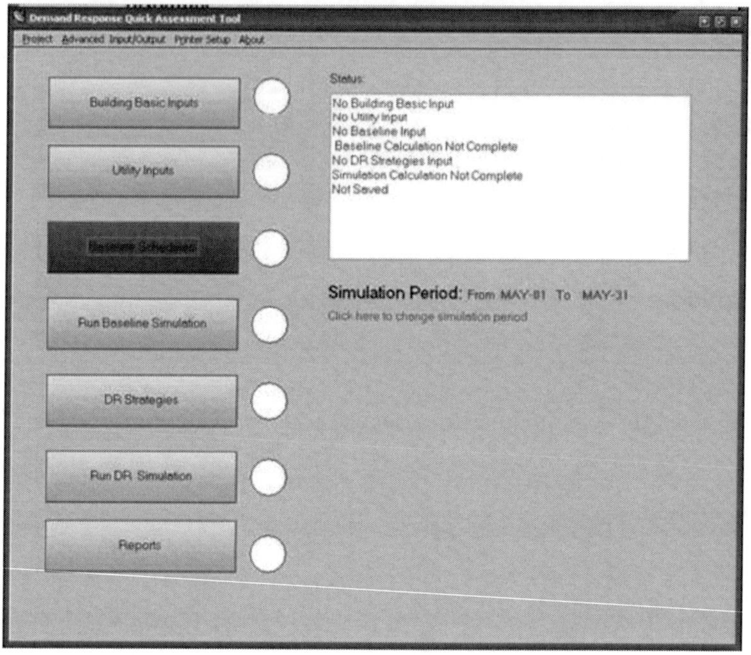

Results of the Demand Response Quick Assessment Tool simulation are presented graphically and textually for analysis.

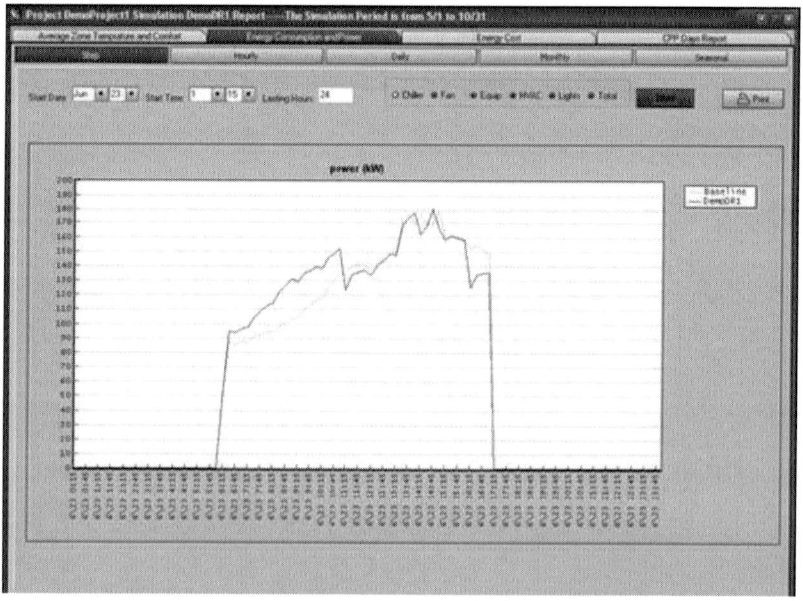

Adding Demand Response

In most electric systems, high penetrations of renewable energy means large amounts of variable wind and solar resources. This translates to a greater need for quick-ramping generators to provide energy when renewable generators cannot. Unfortunately, these quick-ramping generators are often expensive and inefficient. Demand response provides an alternative strategy to meeting demand with high penetrations of renewables. Rather than dispatching elastic supply resources to meet an inelastic demand, demand response makes the demand elastic - shifting it to times when renewable generators are producing power.

In the model, there are two types of demand response: peak-reduction and dynamic oad shifting. Peak reduction is the conventional form of demand response and can be useful in displacing peaking generators. Dynamic load shifting is a more advanced form of demand response that relies on SmartGrid technology to shift flexible loads (e.g. washing machines, dishwashers, electric vehicle charging) to times when renewable energy is abundant. RMI is conducting research to identify how much dynamic load shifting is available for different systems.

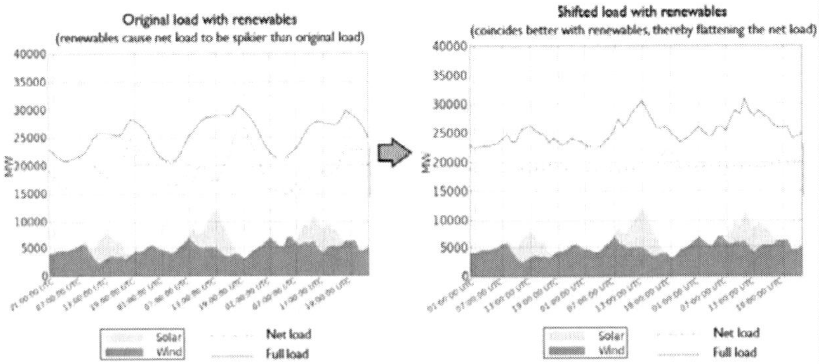

The RMI dispatch model can simulate the capacity mix for alternative scenarios and resources.

Cost-Effectiveness Tools

Determining the cost-effective mix of demand response programs is a central challenge to any demand response provider, regulator, or policymaker. Projections of long-term benefit streams (such as avoided resource costs), identification of relevant costs, and the accurate characterization of these costs are examples of the types of calculations that must be made in assessing a program's cost-effectiveness. The creation of cost-effectiveness tools is essential to determining the most efficient mix of demand response programs for a given service territory or region.

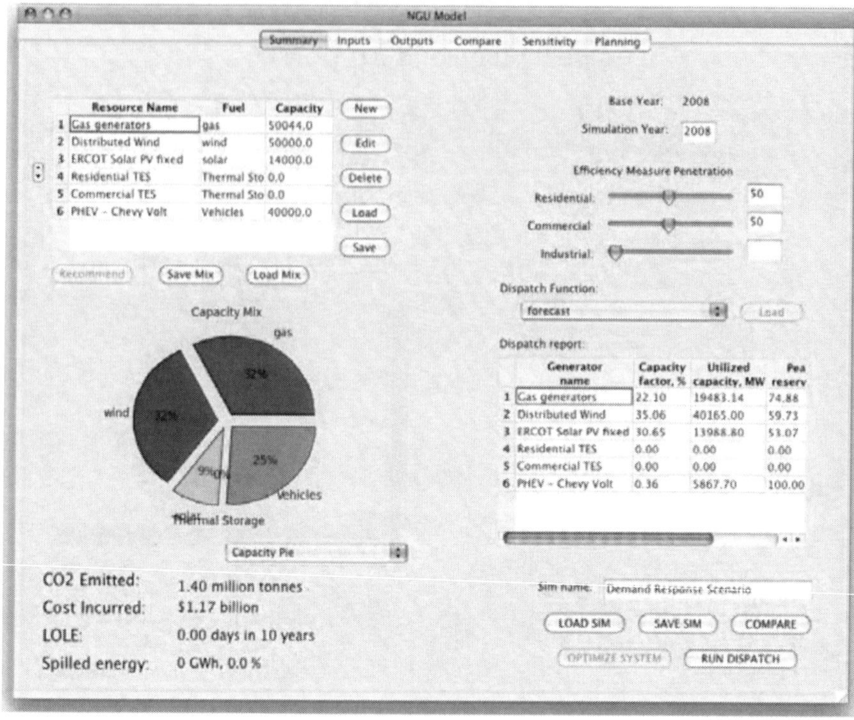

Methods for evaluating the cost-effectiveness of demand-side management programs have been in use for many years. For example, the California Standard Practice Manual provides detailed guidelines for assessing the cost-effectiveness of demand side programs from several different perspectives.[88] The cost-effectiveness tests defined in the report are used by utilities and regulators across the United States and abroad. The federal agency or the Coalition could oversee and support efforts, such as national workshops or further research, to develop new methods or tailor existing methods for analyzing demand response.

A tool to help retail electricity consumers assess the net financial benefits of participating in demand response programs would be valuable, particularly for commercial and industrial consumers. The Demand Response Quick Assessment Tool discussed earlier in this section is a step in this direction. Ideally, such a tool would use the best available information to show customers how their electricity bill would change if they were to participate in a given demand response program.

A final example is a tool that aids load-serving entities, developers of variable resources, and transmission developers in assessing the value and cost-effectiveness of demand response as a resource. This may become especially important in carbon-constrained environments. Case studies of AMI business case filings should be placed in the previously discussed Web-based clearinghouse. Such filings exist for any utility that has sought regulatory approval for an investment in smart meters, and could guide cost-effectiveness analysis in a different region. Beyond case studies, a more detailed tool that would allow the user to customize analysis for a given set of system conditions and cost assumptions would further improve understanding of a given program's potential cost-effectiveness.

Decision Tools for End-Use Consumers

Decision tools are crucial to help end-use customers understand how to change their consumption patterns to achieve bill savings and provide system benefits while enrolled in a demand response program. The federal agency or the Coalition would oversee the identification of new tools that each customer class needs, as well as support the development of selected tools. These tools should draw on past experience and case studies to help customers in specific industries or with specific types of facilities (e.g., warehousing, health care, or retailing) select appropriate demand response providers, programs, and technologies.

Some customers manage multiple facilities, each varying in size and relative location. Examples of such consumers include university campuses, national retailers, and federal or military sites. The development of demand response-related tools and plans to engage these existing customers with special circumstances may help achieve the maximum amount of demand response that can be developed and deployed.

Information in these tools would be updated as market, economic, and regulatory factors change.

Cost Recovery Tools and Methods for Enabling Technology

Tools and methods need to be provided to develop the business case and examine options for recovering costs of installing approved, cost-effective new technologies that enable demand response. Each cost recovery option should include an assessment of the impact on customer rates. Such tools could assist utilities and state and local governing officials that are considering demand response programs by providing important information on the financial impacts of each demand response investment. The federal agency or the Coalition could facilitate discussions between utilities and state and local governing officials so that they can learn best practices from each other with regard to cost recovery.

For example, devices such as advanced meters and smart thermostats are essential for some demand response initiatives and can greatly increase the effectiveness of others. The installation of these devices requires investment from load-serving entities and/or customers. An appropriate method may be needed for utilities or load-serving entities to recover the costs they incur.

Tools and Methods for Incorporating Demand Response in Dispatch, Ancillary Services, Transmission, and Resource Planning

New tools and methods should be developed to directly incorporate demand response into dispatch algorithms and resource planning models. These tools and methods are important to better capture the capability of demand response to serve as an alternative to building new generation and transmission and to act as a resource to alleviate transmission congestion. FERC's Order No. 890 could be instructive in the development of these tools and methods. In Order No. 890, FERC stated that "where demand resources are capable of providing the functions assessed in a transmission planning process, and can be relied upon on a long-term basis, they should be permitted to participate in that process on a comparable basis."[89] The Coalition or federal agency could oversee the development of research and tools examining ways to incorporate demand resources in these processes.

Also needed are better methods to forecast and model the capability of demand resources to adjust consumption in near real-time. Finally, hardware and software tools that allow customer appliances or vehicles to autonomously provide reliability services, such as frequency response service, should be developed.

Transparency of Price Information

For large customers and policymakers to make decisions about the types of economic and business opportunities available in demand response programs, they need access to real-time or near-real-time energy prices. Therefore, there is value in developing tools that provide access to real-time or near-real-time energy prices, ancillary service prices, capacity payments, and congestion costs. Access to transparent price information should also be provided to residential customers when feasible and cost-effective.

To a certain extent, real-time pricing (RTP) programs already provide information to participants about the price of energy. Many utilities have websites that provide information about the next day's hourly prices.

However, additional means are often taken to alert customers about unusually high prices. In Commonwealth Edison's residential RTP program, customers are alerted via e-mail or text message if hourly energy prices are about to exceed a pre-specified threshold. Other dynamic pricing programs will use similar means to convey information to customers about upcoming event days. However, customers do not always use these tools, and some are not aware that they even exist. The federal agency or the Coalition would oversee and support efforts to improve the effectiveness of tool design.

A different example of a tool for providing price information in an innovative way is the Energy Orb. The Energy Orb is a spherical glass lamp that emits different colors to signal off- peak, peak, and critical peak hours of the day. This is a visual aid that enhances a customer's ability to respond to price signals. Several recent dynamic pricing experiments have used the Energy Orb in conjunction with dynamic rates to assess its impacts. The results have been somewhat mixed. In the California Statewide Pricing Pilot, the Energy Orb led to greater conservation and demand response impacts among residential customers.[90] At Baltimore Gas & Electric Company, peak demand reductions increased from 18 to 21 percent without the Energy Orb to 23 to 27 percent when it was provided.[91] However, Connecticut Light & Power's recent pilot had different results: the Energy Orb did not provide any incremental demand response benefits.[92] Regardless, the device serves as an example of a tool that could be used to provide information about price signals to customers.

Table 2 provides a list of the strategies and activities described in the Draft National Action Plan.

SIDEBAR 3: THE DRIVE MODEL

The National Assessment answered an important question about the level of demand response that could be achieved in the United States. It also raised new questions about the implications of achieving this potential. For example:

- How would demand response programs affect load during non-event hours?
- What changes in system operations might demand response require?
- How does demand response affect the need for new capacity?

- How would carbon dioxide emissions change with demand response?
- Would these effects change with the introduction of additional smart grid programs?
- *Ultimately, what is the dollar value (benefit) of these impacts?*

To help answer such questions, FERC staff and The Brattle Group are currently developing the Demand Response Impact and Value Estimation (DRIVE) model. The DRIVE model estimates the impact of a portfolio of demand response and smart grid programs on system operations, using specific characteristics of a power system as inputs.

The DRIVE model is an example of one type of tool that could be created through the efforts of the National Action Plan. A simple user interface makes the model accessible to a broad audience, supporting the educational objectives of the National Action Plan. Its specific focus on demand response and smart grid programs makes it a useful complement to other tools that are available for assessing the system impacts of energy resources.

Preliminary work with the DRIVE model suggests that:

- Modest amounts of demand response can lead to large benefits.
- Avoided capacity cost is a significant benefit.
- Very large reductions in peak demand can lead to the creation of a new peak, which can reduce potential benefits.
- The value of large peak reductions may be enhanced by a broader range of technologies and program offerings, the benefits of which may extend beyond avoided capacity and into other areas like avoided energy costs and reduced carbon dioxide emissions.

Table 2. Strategies and Activities

Requirements for Technical Assistance
• Establish a National Forum • Conduct Informational and Educational Sessions for Policymakers and Regulators • Build a Panel of Demand Response Experts • Sponsor Technical Papers • Establish a Program for Technical Assistance for Demand Response • Provide Demand Response Grants
Requirements for National Communications Program

- Develop a Communications Umbrella
  - Program Development
    - Compile Case Studies
    - Identify Demonstration Program Sites
    - Conduct Program Evaluation
  - Message Framework Development
    - Conduct Foundational Market Research
    - Develop Adaptable, Consumer-Friendly Messages
    - Position Demand Response as Part of Overall Smart Energy Use
    - Develop Communications Materials, Tactics, and Toolkits
    - Develop Support Materials Tailored to Large Commercial and Industrial Customers
- Local Implementation
  - Support for Demonstration Programs
  - Support for Other Demand Response Programs
- Direct Outreach to States, Policymakers, and Partners
  - Communications Assistance to States and Policymakers
  - Partnership Development
- Develop a Web-Based Clearinghouse of Demand Response Materials
  - Case Studies of State Laws, Regulations, and Tariffs
  - Database of Pilots, Programs, and Markets
  - Quantitative and Qualitative Summaries
  - Measuring Program Results with Criteria and Metrics
  - Information for Program Implementation
  - Online Message Board Capability
  - Information About the Development of Standards and Protocols
- Develop or Enhance Demand Response Estimation Tools and Methods
  - System Impact Estimation and Prediction Tools
  - Cost-Effectiveness Tools
  - Decision Tools for End-Use Customers
  - Cost Recovery Tools and Methods for Enabling Technology
  - Tools and Methods for Incorporating Demand Response in Dispatch, Ancillary Services Transmission, and Resource Planning
  - Transparency of Price Information

APPENDIX A : RESEARCH GAPS

In developing a list of possible areas for further research, some potential resource and research gaps have been identified. Research topics to fill these gaps are listed below, organized by whether each topic relates more to social science[93] research or technical issues. It will be the role of the federal agency and the Coalition, working with its members, to prioritize the topics, identify additional topics, and select a subset as candidates for further research.

Social Science Research Issues

- Exploration of issues related to rate design, metering costs, and an analysis of customer response to time-varying prices in successful programs.
- How the participation rate and demand impacts vary depending on whether the demand response rate design is required by law or regulation (e.g., as the default rate) or if customers voluntarily sign up for it.
- Estimates of customer price sensitivity by type and class of customer.
- The short run price elasticities from pilot demand response programs are results that originate in a few states (and Canadian provinces); investigate whether these results are applicable to other areas, and also investigate how results would change over the long run (i.e., whether they would be higher because of equipment purchase decisions favoring storage and battery technologies or lower because of lack of persistence).
- Investigation of typical demand response participation rates and how these rates vary based on program design parameters, marketing strategy, customer type (residential, commercial, and industrial; gender and income level), and customer profile (e.g., age, geographic location). Also, investigation into the non-price response incentives that motivate customers to participate in demand response programs and products.
- Exploration of the effectiveness of demand response when critical days are sustained for multiple days in a row and the persistence of demand response behavior when programs and products are implemented over a long-term time horizon.
- Understanding how consumers shift demand to off-peak hours of the day and if consumers take conservation measures to lower peak demand over all hours of the day, not just during peak hours.
- How demand response programs compare with energy efficiency programs in terms of cost-effectiveness and demand reduction efficacy.

- Identification of the most cost-effective demand response programs for customers who do not have advanced metering infrastructure.
- Impacts of customers participating in multiple demand response programs.
- Research on electricity use by home office workers.
- Research on effective gender-based, age-based, and race/ ethnicity-based messages.
- Research to identify "discretionary use" appliances — i.e., appliances whose operations can easily be scheduled for off-peak times.
- Usage of environmentally oriented feedback information, such as carbon content, to motivate behavior change.
- Impacts and benefits to low-income consumers.

Technical Research Issues

- Relative benefits and costs of various types of demand response programs, products, technologies, and incentives, covering such topics as the various types of benefits, potential costs, and payback horizons associated with each.
- Better understanding of the optimal amount of consumer versus demand response provider control of appliances.
- An analysis of the environmental effects of demand response, such as: an estimate of emissions mitigation when considering load reduction versus load shifting; an assessment of the potential for demand response to assist in the integration of variable resources such as renewable energy; and a calculation of environmental benefits stemming from the reduced need to build new generation capacity and transmission systems.
- A study of how long demand response resources can be expected to provide resource adequacy and reliability benefits compared with other resources, such as generation, transmission, and storage.
- A study of how demand response resources can be dispatched to support and balance variable generation from renewable energy.
- A study of how plug-in hybrid electric vehicles interact with demand response programs, examining whether demand response rate design provides a price signal that encourages PHEVs to charge during off-

peak hours, as well as how different demand response pricing mechanisms interact with PHEVs and their net impacts on how demand for electricity might change.

- Research into the appropriate codes and standards (e.g., building codes, consistent definitions, performance standards) for demand response, considering whether they will be mandated at the federal or state level; whether they are mandatory or voluntary requirements; how they will be complied with and enforced; and how the costs involved in their development, compliance, and enforcement will be paid for. Integration of the many utility and third-party programs, devices, and software applications needs to be taken into consideration.
- Advanced metering topics such as estimating initial costs, methods used to recover costs, calculating time to payback, the functionality of the devices, and dealing with swift technology advancement.
- Smart metering for residential customers, covering topics such as categories and characteristics of various smart metering systems, cost-effectiveness, setting up the communications and database, customer awareness, access to data, retrofitting existing meter connections with smart technology, and consumer protections.
- Implementation of demand response programs and products at government-owned campuses, covering topics such as the types of demand response suitable for government customer applications, barriers and lessons learned from existing programs, percentages of the state electricity portfolio under municipal, county, state, and federal control, and the potential impact of incorporating government loads into demand response programs and products.
- Ensuring smart grid cybersecurity, system reliability, and systems interoperability, especially considering the massive amounts of data that could be transferred between customers and utilities and other demand response providers.
- An assessment of the job skills and training needed to successfully deploy demand response programs, products, technologies, and incentives.
- Best strategies to coordinate retail demand response programs with RTO/ISO demand response programs for organized power markets.
- A study of the benefits of a national demand response policy, including reliability, efficiency, economic and environmental benefits,

and benefits related to consumer savings; the increased ability to integrate variable generation; and the increase in potential competitive sources of ancillary services.

- Economic and reliability impacts of PHEVs as they penetrate the market.
- Exploration of the effectiveness of various types of in-home displays.
- Exploration of impacts that the battery life of various home area network devices and in-home displays will have on residential participation in demand response programs.
- Exploration of impacts on appliance/electronics longevity that are cycled on/off by demand response programs.
- Research consumer willingness to sell back energy from micro-generation and home storage to offset peak periods.
- Best strategies to maximize deployment of demand response potential for states with specific programs or needs, taking into account electric demand profile, generation mix, preferred type of demand response programs and products, regional trends, stakeholder views, and previous assessment research.
- Curtailable load programs aimed at industrial customers, covering topics such as potential benefits to all customers, cost recovery, rate design, and program marketing.
- The relationship of demand response to other state energy initiatives (e.g., energy efficiency, renewable energy, smart grid). This research would explore how demand response overlaps or complements these other initiatives and where management and oversight could be leveraged or combined.
- Understanding the interest of elected officials, regulatory agencies, utilities and load-serving entities, RTOs/ISOs, other demand response providers, and third-party vendors in making demand response a priority.
- The ability to integrate large-scale renewable energy resources and electric vehicles, which is critical to advance the benefits of demand response.
- The impact of distributed generation as a resource that can be integrated more effectively in conjunction with demand response.
- An assessment of the location impacts of demand response.
- A study that identifies the load and peak demand profiles of the electricity control areas.

Appendix B : Examples of Demand Response Programs

This appendix presents a few examples of demand response retail and wholesale programs around the United States. To put these case studies in context, Figure 5, which is taken from the National Assessment, shows the regional variation in the national potential for demand response.

Retail Programs

Arizona

Arizona is a state with vertically integrated investor-owned and publicly owned utilities. It has very hot summers that bring with them a large reliance on central air conditioning. To curtail growth in peak loads, the two major utilities — the Arizona Public Service Company and the Salt River Project — have offered time-of-use (TOU) pricing programs to their residential customers for more than two decades.[94] During this time, they have attracted some 30 to 40 percent of the residential market.[95] For example, the Salt River Project has more than 222,000 TOU customers, or about 25 percent of its electric customers, which is the third-largest TOU initiative in the United States.[96] Typical TOU customers on this program lower their bill by 7 percent.[97] What is noteworthy is that these effects were obtained without the intervention of enabling technologies.

This case study indicates that even without enabling technologies, significant customer participation can be achieved in time-based rates if they are well-designed and marketed, and that these rates can successfully lower peak demand for many years.

California

California is a leader in exploring and expanding demand response. For instance, the California Energy Commission issued regulations instituting mandatory TOU pricing for large commercial and industrial customers in 1978. In addition, regulations for appliance and building energy consumption[98] are a major reason that per capita electricity use in California has stayed constant during the past three decades, while the per capita consumption in the United States has grown appreciably.[99] Later, during the western energy crisis of 2000–2001, California developed an Energy Action Plan that specified a loading order of resources: first, energy efficiency; second, demand response; third, renewable energy sources; and finally, conventional generation options.[100]

The Energy Action Plan also set a target reduction in peak demand of 5 percent for 2007.[101] When California's three investor-owned electric utilities only achieved 40 percent of that target in 2007, they were asked to revamp their efforts. Simultaneously, the California Public Utilities Commission (CPUC) approved business cases for the deployment of advanced metering initiatives (AMI) by all three utilities[102] and then issued a landmark ruling making dynamic pricing the default tariff for all nonresidential customers who were part of AMI. In addition, California operates the "Flex Your Power" campaign to promote customer awareness of pending energy shortages and to encourage them to curtail energy use.[103] This program has achieved very high levels of brand recognition because it is featured in a number of prominent places such as billboards, train stations, and websites, as well as on postmarks.

California also carried out the nation's first comprehensive test with dynamic pricing, known as the Statewide Pricing Pilot, which involved approximately 2,500 residential and small commercial and industrial customers.[104] The results of the Statewide Pricing Pilot provided information about the likely amount that customers would lower their peak demand at different price levels, both with and without enabling technologies.[105] In addition, a large number of participants elected to continue on the experimental rates despite a new metering charge, indicating that participants were satisfied with the experimental rates.[106] Overall, the experiment showed that if customers are fully informed, they are more likely to participate in dynamic pricing programs.

This case study indicates that if a state makes a strong commitment to dynamic pricing, it can achieve a significant reduction in peak demand; that enabling technologies can further boost impacts; and that by making demand response the default option, large numbers of customers can be induced to participate in demand response.

Connecticut

The Connecticut legislature passed legislation that required electric distribution companies to conduct integrated resource planning beginning on January 1, 2008.[107] The identified resource needs must first be met through all available energy efficiency and demand reduction resources that are cost-effective, reliable, and feasible.[108] Among other things, the electric distribution companies' procurement plans must specify the amount of demand-side measures, such as demand response, that can meet the capacity planning needs.[109]

In summer 2009, Connecticut Light & Power ran a pilot program for demand response to discover how to most effectively reduce electricity usage during peak periods when demand for electricity reaches a critical level.[110] Data from this program have been analyzed, and the results have led to a number of conclusions about customer response to prices, both with and without various enabling technologies.[111]

This case study indicates that legislation can be passed that encourages the use of demand response. It also shows how pilot programs are being implemented.

Delaware

Legislation in Delaware requires Delaware Power & Light to conduct integrated resource planning, commencing December 1, 2006.[112] This biennial resource report requires Delaware Power & Light to project a 10-year supply and demand forecast. Additionally, Delaware Power & Light must propose the resource mix to meet its long-term obligations over the 10- year period, including the use of demand-side management programs.[113]

This case study is an example of legislation that has been passed to encourage the use of demand response for integrated resource planning.

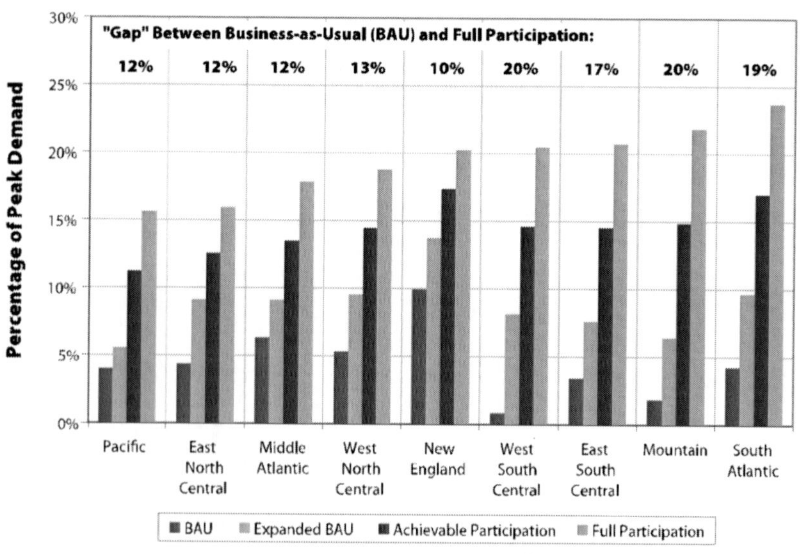

Figure 5. Regional Demand Response Potential.

Florida

In the panhandle of Florida, Gulf Power, a subsidiary of Southern Company, has been operating the Energy Select (formerly known as the Good Cents SELECT program) since the early 1990s.[114] This program includes an automated energy management system for homes that comes with a critical-peak pricing (CPP) tariff. The average rate is around 8 cents per kilowatt-hour, and the CPP tariff applied during the top 1 percent of the hours is four times higher.[115] The system controls four end-uses: air conditioning, space heating, water heating, and swimming pool pumps. Some 8,000 customers subscribe to the program and save up to 15 percent on their monthly energy bills after paying a monthly fee of about $5 a month.[116] Gulf Power estimates that the program induces a drop of 2 kW per participating customer, amounting to approximately 40 to 50 percent of customer load during the top 1 percent of the hours of the year.[117] The program is generating interest among other Florida utilities as well. One utility is running a pilot to test customer acceptance and response. The pilot is being closely tracked by the other utilities.

Florida Power & Light's (FPL's) demand response program, "On Call," is one of the largest load management systems (LMSs) in the nation. At the end of 2008, FPL's goal was to provide 973 MW of load control during times of high demand on its system. The system uses more than 900,000 load control transponders connecting more than 780,000 users, and the load control capability is 984 MW. FPL uses a power line communications system that offers many control strategies enabled by a two-way communications feature. When managed effectively, this system can be an attractive economic alternative when compared with the total cost of adding new peak load power generating plants.

An essential element of the success of the program is keeping customers enrolled in the program and maintaining the equipment in the field to ensure that the system delivers the demand reduction that is expected by the system operators. About 1.5 percent of the customers who have the LMS field equipment installed in their home call the customer care center with questions. These calls can be as simple as a request to add additional appliances to the load control program, although in other cases, customer service representatives may have to respond to a customer's perception that their participating appliances failed because of the installed LMS equipment. Employees specially trained in LMS (including a basic electrical knowledge of air conditioning, water heaters, and pool pump operation/wiring) help reduce customer dropouts from the system and ensure continued customer participation.

FPL customers benefit from the On Call load control program because it allows FPL to maintain a consistent supply of power generation during times of peak demand. The demand reduction enabled by On Call customers has helped keep electric rates low by deferring the need to build three power plants since the program's inception, which also helps protect the environment.[118]

This case study indicates that dynamic pricing coupled with enabling technologies can bring about substantial savings to both the utility and the customer.

Georgia

Since the mid-1990s, Georgia Power has operated the nation's largest real-time pricing program directed at large commercial and industrial customers.[119] Its current subscribers include approximately 1,200 medium to large industrial and commercial customers (greater than 250 kW), amounting to more than 4,000 MW of summer peak demand.[120] Most of the customers participate in the day-ahead program, in which they receive information on the next day's hourly prices on the afternoon of the previous day. About 100 of the largest industrial customers participate in an hour-ahead program, in which customers receive a day-ahead forecast of hourly prices, but only an hour's notice of the firm price in the next hour. Customers are offered a variety of price protection products to shield them from price volatility. They can chose the combination of risk exposure that best matches their attitudes toward risk and their ability to adjust their operations in response to variations in the price of power. The maximum load reductions on the moderate price day are approximately 4 percent for day-ahead customers and 10 percent for hour-ahead customers. On the high- price day, estimated load reductions rise to 7 percent for day-ahead customers and 30 percent for hour-ahead customers.[121]

This case study indicates that real-time pricing is feasible and cost-effective for medium and large commercial and industrial customers and that well-designed programs can recruit large numbers of participants to real-time pricing in this market segment. It also shows that price protection and other risk management products may feasibly be included in demand response programs.

Illinois

In June 2006, Illinois legislation enabled real-time pricing.[122] Each electric utility servicing at least 100,000 customers was required to file a tariff allowing residential customers to elect real-time pricing by January 2007. In setting the tariff, each electric utility was required, at a minimum, to describe its methodology and the manner in which real-time customers would have access to hourly and day-ahead prices.

To implement real-time pricing, each electric utility was required to install a meter that is capable of recording hourly intervals of energy use for each customer that elects real-time pricing. In addition, each utility was entitled to recover incurred costs in implementing real-time pricing. These costs were imposed on all customers — those who elected to participate in a real-time pricing rate and those who did not.

At the end of 2007, the first full year required under the law, Commonwealth Edison had 3,334 and Ameren had 500 active participants in their respective residential real-time pricing programs. In that year, average bill savings for all participants in the Commonwealth Edison program was about 13 percent.[123] Ameren participants, on the other hand, had a total savings of 16 percent compared with what the same bill would have been under the flat rate.[124]

By the end of 2008, Commonwealth Edison and Ameren had increased their enrollments to 5,838 and 3,147 active participants, respectively, though average bill savings fell to 2.5 percent for all participants in the Commonwealth Edison program[125] and 9 percent in the Ameren program.[126]

This case study indicates that legislation can be used to encourage dynamic pricing. Also, real-time pricing rates are feasible for residential customers, and they can achieve significant savings in the residential market.

Massachusetts

In July 2008, Massachusetts Governor Deval Patrick signed the Green Communities Act, a comprehensive energy reform bill.[127] Among its many features, this law makes energy efficiency programs compete in the market with traditional energy supply. Utility companies are now required to purchase all available energy efficiency improvements that cost less than it does to generate power. Additionally, utility companies are now required to offer rebates and other incentives for customers to upgrade lighting, air conditioning, and industrial equipment to more efficient models whenever those incentives cost less than generating the electricity it would take to power older, less-efficient equipment.[128]

Michigan

In January 2007, the Michigan Public Service Commission (PSC) submitted its *21st Century Electric Energy Plan*[129] in response to an executive order from the governor the year before. One of the recommendations in the energy plan is that Michigan's PSC be authorized to require the use of active load management measures by utilities (forecasted to reduce demand by 570 MW in 10 years). The plan also recommends pilot programs employing advanced metering infrastructure to provide real-time price information to customers.

In Consumers Energy's Balanced Energy Initiative, the company forecasted that load management, demand response, and interruptible load programs would reduce retail peak by 9.3 percent by 2030.[130] The reduction would be implemented through a comprehensive AMI program that includes central air conditioning load management and demand response programs for residential and small commercial customers. This program would reduce peak demand as well as ongoing energy requirements.[131]

Consumers Energy plans to conduct a demand response and information pilot aimed at assessing customer behavior in the smart grid environment. The centerpiece of this pilot is a randomized program that places AMI-enabled customers into various dynamic pricing and information treatments while comparing them with strategic control groups.

The research is distinguishable from previous demand response pilot programs because it will 1) carefully use treatment randomization to achieve internal and external validity and 2) address a set of questions that extends beyond a standard investigation of load-shifting behavior.[132] At a high level, this research aims to:

- Understand how dynamic pricing can best be incorporated into Consumers Energy's portfolio.
- Understand consumer responses to dynamic electricity pricing, during both peak and off-peak periods (to examine demand shifting vs. demand reduction).
- Evaluate whether automated demand response technology enhances consumer responses to dynamic pricing.
- Compare consumer response to CPP against critical peak rebates.
- Understand consumer responses to real-time provision of information regarding electricity usage, during both peak and off-peak periods.

- Evaluate potential climate change mitigation impacts of demand response programs and information provision.
- Evaluate the extent to which participation in demand response and/or information programs motivates lasting energy efficiency improvements.
- Evaluate the heterogeneity of responses to these programs across subpopulations.

While the CPP and real-time pricing programs may appeal to many of Consumers Energy's customers, a portion of their customer base may prefer a different, potentially less involved program that can provide them with opportunities to reduce their bills while requiring minimal action on their part. To provide these customers with a program of that nature, Consumers Energy will concurrently roll out Load Management and advanced Direct Load Control programs that will leverage AMI investment.[133]

In addition to future efforts, demand reduction has been conducted in Michigan for many years. Detroit Edison maintains a significant Direct Load Control Interruptible Air Conditioning (IAC)[134] program, which was established more than 20 years ago during the days of demand-side management. IAC has grown to more than 280,000 enrolled customers (of 1.9 million total customers, with a 79.1 percent central air conditioning saturation)[135] who receive a $0.02 rate break during the cooling months of June–October in exchange for the possibility of interruption by the company via remote control relays. This pricing structure is unique because it provides a discount on the hourly rate instead of a fixed amount monthly credit.

The current diversified per customer load reduction is 0.85 kW, which is down considerably from approximately 1.2 kW 10 years ago. This decrease in load reduction is due to the increase in efficiency of central air conditioning units over the years.[136]

The IAC program is intended to help curtail peak load during emergency or high price periods. It is also utilized as capacity reserve, as it is available within 10 minutes. The program is a cycling program at 15-minute on/off intervals that is capable of shedding up to 230 MW of load on days over 90 degrees Fahrenheit. IAC may be cycled for no more than 8 hours within a 24-hour period.[137]

In addition, the program is used for localized substation load relief. Detroit Edison developed a remote control program using IAC customers (identified by the substation serving them) that is capable of interrupting only customers taking service within substation-serviced areas.[138] This program has proven very successful at relieving load on overloaded substations and avoiding localized circuit rotation, brownouts, and, in extreme cases, the loss of a substation transformer.[139]

This case study is an example of legislation being passed to encourage new types of demand response programs to complement existing, successful programs.

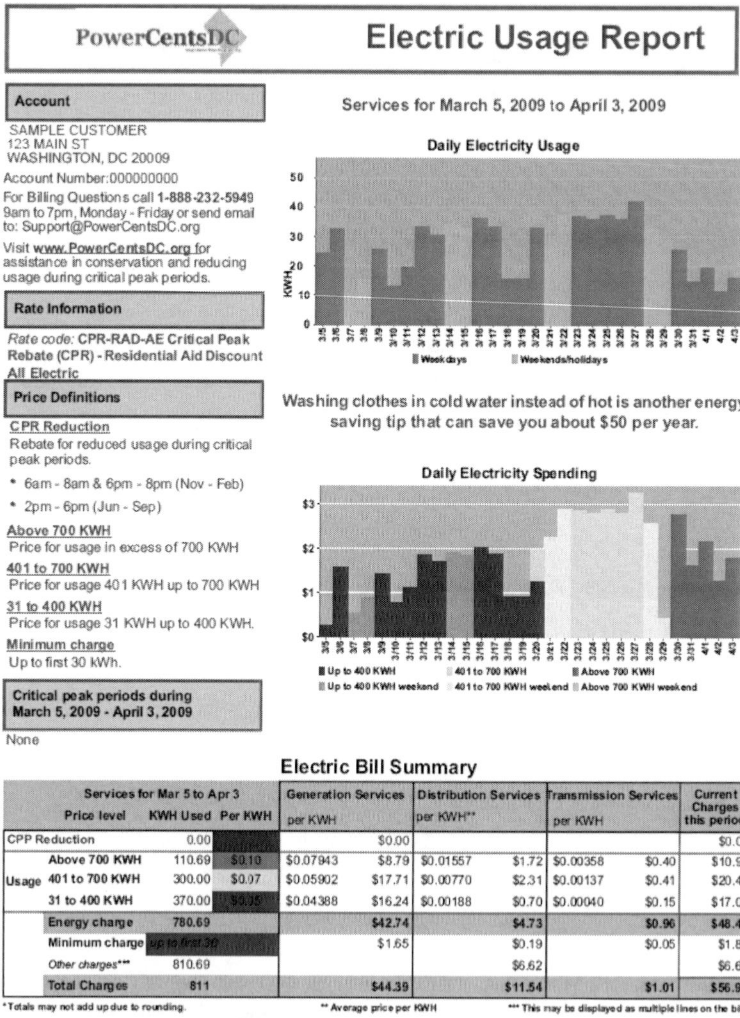

PowerCentsDC **Electric Usage Report**

Account

SAMPLE CUSTOMER
123 MAIN ST
WASHINGTON, DC 20009

Account Number:000000000

For Billing Questions call **1-888-232-5949** 9am to 7pm, Monday - Friday or send email to: Support@PowerCentsDC.org

Visit www.PowerCentsDC.org for assistance in conservation and reducing usage during critical peak periods.

Rate Information

Rate code: **CPR-RAD-AE Critical Peak Rebate (CPR) - Residential Aid Discount All Electric**

Price Definitions

CPR Reduction
Rebate for reduced usage during critical peak periods.

- 6am - 8am & 6pm - 8pm (Nov - Feb)
- 2pm - 6pm (Jun - Sep)

Above 700 KWH
Price for usage in excess of 700 KWH

401 to 700 KWH
Price for usage 401 KWH up to 700 KWH

31 to 400 KWH
Price for usage 31 KWH up to 400 KWH.

Minimum charge
Up to first 30 kWh.

Critical peak periods during March 5, 2009 - April 3, 2009

None

Services for March 5, 2009 to April 3, 2009

Washing clothes in cold water instead of hot is another energy saving tip that can save you about $50 per year.

Electric Bill Summary

Services for Mar 5 to Apr 3	Price level	KWH Used	Per KWH	Generation Services per KWH		Distribution Services per KWH**		Transmission Services per KWH		Current Charges this period
CPP Reduction		0.00			$0.00					$0.00
Usage	Above 700 KWH	110.69	$0.10	$0.07943	$8.79	$0.01557	$1.72	$0.00358	$0.40	$10.91
	401 to 700 KWH	300.00	$0.07	$0.05902	$17.71	$0.00770	$2.31	$0.00137	$0.41	$20.43
	31 to 400 KWH	370.00	$0.05	$0.04388	$16.24	$0.00188	$0.70	$0.00040	$0.15	$17.09
	Energy charge	780.69			$42.74		$4.73		$0.96	$48.43
	Minimum charge	up to first 30			$1.65		$0.19		$0.05	$1.89
	*Other charges****	810.69					$6.62			$6.62
	Total Charges	811			$44.39		$11.54		$1.01	$56.94

*Totals may not add up due to rounding. ** Average price per KWH *** This may be displayed as multiple lines on the bill.

Figure 6. PowerCentsDC Pilot Program Sample Bill.

Washington, D.C.

It is very expensive and time-consuming to change billing systems and processes for sending bills to customers. Instead of overhauling the entire billing system and the process of mailing bills, it is possible to include a simple dynamic pricing line item adjustment in the current billing system and provide a detailed explanation of this dynamic pricing adjustment in a bill insert. An example of such a demand response bill insert, provided from Washington's PowerCentsDC pilot program,[140] is provided in Figure 6. This example is not the actual bill that pilot program participants receive. Rather, this is supplemental information included as an insert with the actual bill.[141]

This case study provides an example of a billing structure that considers the many regulatory and information technology factors surrounding the billing of customers.

Wholesale Programs

RTO and ISO program offerings are summarized in Table 3.[142]

Table 3. Wholesale Programs

Name	Service Type	Trigger	Response Time
California ISO			
Participating Load Program	Energy	Market bid (energy price > offer price)	1 hour
Participating Load Program	Reserve	Market bid (capacity bid and separate energy bid > offer price)	10 minutes
Electric Reliability Council of Texas			
Emergency Interruptible Load Service	Capacity	Reliability	10 minutes
Loads Acting as a Resource Providing Responsive Reserve Service — Under Frequency Relay Type	Reserve	Reliability (automatic relay)	0.5 seconds (automatic relay) or 10 minutes (phone)
Loads Acting as a Resource Providing Responsive Reserve Service — Controllable Load Resource Type	Reserve	Reliability	Continuous, similar to governor action by a generator; and 10-minute response for remaining obligation to electronic instruction

Table 3. (Continued)

Name	Service Type	Trigger	Response Time
Loads Acting as a Resource Providing Non- Spinning Reserve Service	Reserve	Reliability	30 minutes
Controllable Load Resources Providing Regulation Service	Regulation	Reliability	Effectively instantaneous
ISO-New England			
Real Time Demand Response Program (Capacity Component)	Capacity	Reliability	10 minutes/30 minutes
Real Time Demand Response Program (Energy Component)	Energy	Reliability	10 minutes/30 minutes
Day-Ahead Load Response Program for RTDRP	Energy	Market bid (day-ahead LMP = or > offer price)	Effectively instantaneous
Day-Ahead Load Response Program for RTPR	Energy	Market bid (day-ahead LMP = or > offer price)	Effectively instantaneous
Demand Response Reserves Pilot	Reserve	Reliability	30 minutes
Real Time Price Response Program	Energy	Market bid (day-ahead or forecast real-time LMP = or > $100/MWh)	Effectively instantaneous
Real Time Demand Response Resource	Capacity	Reliability	30 minutes
FCM: On-Peak, Seasonal Peak Resources	Capacity	Reliability	Effectively instantaneous
Real Time Emergency Generation Resource	Capacity	Reliability	Effectively instantaneous
Midwest ISO			
Emergency Demand Response	Energy	Reliability	Resource-specific (biddable parameter)
Demand Response Resource Type I	Energy	Market bid (energy price > offer price)	5 minutes
Demand Response Resource Type I	Reserve	Market bid (energy price > offer price)	10 minutes
Demand Response Resource Type II	Energy	Market bid (energy price > offer price)	5 minutes
Demand Response Resource Type II	Reserve	Market bid (energy price > offer price)	10 minutes

Name	Service Type	Trigger	Response Time
Demand Response Resource Type II	Regulation	Market bid (energy price > offer price)	Effectively instantaneous
Load Modifying Resource	Capacity	Reliability	—
New York ISO			
Day-Ahead Demand Response Program	Energy	Market bid (energy price > offer price) (security-constrained unit commitment)	—
Demand Side Ancillary Services Program	Reserve	Market bid (energy price > offer price) (security-constrained economic dispatch) (10 minute ramp)	10 minutes
Demand Side Ancillary Services Program	Reserve	Market bid (energy price > offer price) (security-constrained economic dispatch) (10 minute/30 minute ramp)	10 minutes/30 minutes
Demand Side Ancillary Services Program	Regulation	Market bid (energy price > offer price) (security-constrained economic dispatch)	Effectively instantaneous
Emergency Demand Response Program	Energy	Reliability	2 hours
Installed Capacity Special Case Resources (Energy Component)	Energy	Reliability	2 hours
Installed Capacity Special Case Resources (Capacity Component)	Capacity	Reliability	2 hours
PJM			
Economic Load Response	Energy	Market bid (self-scheduled, cleared day-ahead bid, or real-time dispatch)	Resource-specific
Economic Load Response	Reserve	Reliability (one hour notice)	10 minutes
Economic Load Response	Reserve	Reliability (day-ahead notice)	30 minutes
Economic Load Response	Regulation	Reliability	Effectively instantaneous
Emergency Load Response — Energy Only	Energy	Reliability	1 hour or 2 hours (participant-selected)

Table 3. (Continued)

Name	Service Type	Trigger	Response Time
Full Emergency Load Response (Capacity Component)	Capacity	Reliability	1 hour or 2 hours (participant-selected)
Full Emergency Load Response (Energy Component)	Energy	Reliability	1 hour or 2 hours (participant-selected)
Southwest Power Pool			
Variable Dispatch Demand Response	Energy	Market bid (energy price > offer price) (security-constrained economic dispatch)	5 minutes

FCM: forward capacity market
LMP: locational marginal price
MWh: megawatt-hour
RTDRP: real-time demand response program
RTPR: real-time price response

APPENDIX C: LIST OF COMMENTERS

The following commenters presented comments at the Technical Conference held at the Commission's headquarters on November 19–20, 2009:

- Wal-Mart Stores, Inc., Ken Baker
- Illinois Commerce Commission, Sherman Elliott
- Florida Power & Light, John Haney
- Demand Response Coordinating Committee, Dan Delurey
- Sacramento Municipal Utility District, Jim Parks
- Hess Corporation, Jay Kooper
- National Rural Electric Cooperative Association, David Mohre
- Ohio Consumers' Counsel, Janine Migden-Ostrander
- Alcoa, Inc., Walter Brockway
- ISO-New England, Inc., Henry Yoshimura
- Energy Curtailment Specialists, Paul Tyno
- To the Point, Judith Schwartz
- General Electric, Mary Bell
- IBM Energy and Utilities, Don Mak

The following submitted written comments to the Commission in response to the Discussion Draft:

- American Public Power Association
- Brett Feldman
- California Department of Water Resources State Water Project
- California Public Utilities Commission
- Californians for Renewable Energy, Inc.
- Constellation Energy Commodities Group, Inc. and Constellation NewEnergy, Inc.
- Demand Response and Smart Grid Coalition
- Demand Response Coordinating Committee
- District of Columbia Public Service Commission
- Edison Electric Institute
- Electricity Consumers Resource Council
- Energy Curtailment Specialists, Inc.
- Environmental Defense Fund
- Federal Trade Commission
- Hess Corporation
- heyCoop, LLC
- Ice Energy, Inc.
- National Association of Regulatory Utility Commissioners
- National Association of State Utility Consumer Advocates
- National Grid USA
- National Rural Electric Cooperative Association
- New York State Public Service Commission
- North America Power Partners LLC
- North American Electric Reliability Corporation
- Occidental Chemical Corp. et al.
- Ohio Public Utilities Commission
- Pacific Gas & Electric Co.
- Peak Load Management Alliance (Aztech Associates, Inc., Comverge, Inc., CPower, Inc., Energy Curtailment Specialists, Schneider Electric, Southern California Edison, and Summit Blue Consulting)
- Portland General Electric Co.
- Project for Sustainable FERC Energy Policy
- PSEG Companies

- Robert Borlick
- San Diego Gas & Electric Co.
- Smart Grid Library (Christine Hertzog)
- Southern California Edison Co.
- Steel Producers
- Tendril Networks, Inc.
- To the Point
- Utilimetrics
- Utility Economic Engineers (Mark B. Lively)
- Wal-Mart Stores, Inc.
- Ziphany, LLC

The following submitted written comments to the Commission in response to the Draft National Action Plan:

- American Electric Power Service Corp.
- American Public Power Association
- Association of Home Appliance Manufacturers
- California Public Utilities Commission
- Demand Response Coordinating Committee
- Demand Response and Smart Grid Coalition
- Edison Electric Institute
- Electric Power Supply Association
- Energy Curtailment Specialists
- EnerNOC, Inc.
- GE Energy
- Gridwise Architectural Council
- Ice Energy, Inc.
- Illinois Commerce Commission
- Indicated New York Transmission owners
- ISO New England
- Judith Schwartz
- Maine Public Utilities Commission
- Maryland Public Service Commission
- Massachusetts Department of Public Utilities
- Midwest Independent System Operator, Inc.
- North American Energy Standards Board

- National Association of Regulatory Utility Commissioners
- National Association of State Utility Consumer Advocates
- National Energy Marketers Association
- National Grid
- National Rural Electric Cooperative Association
- New York Independent System Operator
- Public Interest Organizations/Project for Sustainable FERC Energy Policy
- San Diego Gas & Electric Co.
- Southern California Edison Co.
- Southern Company
- Tendril Networks, Inc.
- Wal-Mart Stores, Inc.
- Xcel Energy Services, Inc.

End Notes

[1] Energy Independence and Security Act of 2007, Pub. L. No. 110-140, § 529, 121 Stat. 1492, 1664 (2007) (to be codified at National Energy Conservation Policy Act, 42 U.S.C. §§ 8241, 8279).

[2] *Discussion Draft on Possible Elements for the National Action Plan* (Oct. 2008), *available at* http://www.ferc.gov/EventCalendar/Files/ 20091028124306-AD09-10-000-Discussion.pdf (Discussion Draft).

[3] Docket No. AD09-10.

[4] *Draft for Comment of the National Action Plan on Demand Response* (March 2010), *available at* http://www.ferc.gov/legal/staff-reports/03-12-10-demand-response.pdf (Draft National Action Plan).

[5] Local demand response implementers are entities, such as state or local governing officials, utilities, load-serving entities, and other demand response providers, interested in providing demand response programs, services, rates, incentives, etc., at a local level.

[6] *A National Assessment of Demand Response Potential* (June 2009), *available at* http://www.ferc.gov/legal/staff-reports/06-09-demand-response.pdf (National Assessment).

[7] Energy Independence and Security Act of 2007, Pub. L. No. 110-140, § 529, 121 Stat. 1492, 1664 (2007) (to be codified at National Energy Conservation Policy Act, 42 U.S.C. §§ 8241, 8279).

[8] *Discussion Draft on Possible Elements for the National Action Plan* (Oct. 2009), *available at* http://www.ferc.gov/EventCalendar/Files/ 20091028124306-AD09-10-000-Discussion.pdf (Discussion Draft).

[9] Docket No. AD09-10.

[10] *Draft for Comment of the National Action Plan on Demand Response* (March 2010), *available at* http://www.ferc.gov/legal/staff-reports/03-12-10-demand-response.pdf (Draft National Action Plan).

[11] EISA section 529.

[12] National Assessment, *supra* note 6.

[13] EISA section 529(c).

[14] *See Wholesale Competition in Regions with Organized Electric Markets*, Order No. 719, FERC Stats. & Regs. ¶ 31,281 (2008), *order on reh'g*, Order No. 719-A, 74 Fed. Reg. 37,776 (Jul. 29, 2009), FERC Stats. & Regs. ¶ 31,292 (2009), *order on reh'g*, Order No. 719-B, 129 FERC ¶ 61,252 (2009).

[15] Time-of-use rates are not dynamic (because the rates for each time period are fixed and do not change with system conditions) and were not included in the National Assessment as a demand response program. However, they are being included here as an element of the expanded array of opportunities for customers to reduce demand in response to signals.

[16] *Smart Grid Policy*, Proposed Policy Statement and Action Plan, 126 FERC ¶ 61,253, at P 1 (2009).

[17] National Assessment, *supra* note 6, at 27.

[18] FERC Staff, *Assessment of Demand Response and Advanced Metering* (2008), *available at* http://www.ferc.gov/legal/staff-reports/12-08-demand-response.pdf.

[19] National Assessment, *supra* note 6, at 42–44.

[20] *Id.* at 27.

[21] *Id.*

[22] Office of Electricity Delivery and Energy Reliability at DOE, *Smart Grid Investment Grant Awards,* http://www.oe.energy.gov/recovery/1249.htm.

[23] The National Assessment provides a state-level analysis of demand response, including a snapshot of existing programs. The full set of results at the state level for 2009 through 2019 is contained within the model developed for the National Assessment, *available at* http://www.ferc.gov/industries/electric/indus-act/demand-response/NADR-model.xls.

[24] *Id.* at 27.

[25] EISA section 529.

[26] EISA section 529.

[27] *Id.*

[28] A transcript of this conference, along with materials provided by speakers, is available on the FERC website at http://www.ferc.gov/industries/electric/indus-act/demand-response.asp.

[29] These comments are available in Docket No. AD09-10, which is the docket for the National Action Plan on Demand Response.

[30] Local demand response implementers are entities, such as state or local governing officials, utilities, load-serving entities, and other demand response providers, interested in providing demand response programs, services, rates, incentives, etc., at a local level.

[31] Larry Cohen et al., *Developing Effective Coalitions: An Eight Step Guide*, in Community Health Education & Promotion: A Guide to Program Design and Evaluation, 144-161 (Mary Ellen Wurzbach ed., 2d ed. 2002).

[32] National Action Plan for Energy Efficiency, *National Action Plan for Energy Efficiency* (2006), *available at* http://www.epa.gov/cleanenergy/documents/napee/napee_report.pdf.

[33] National Action Plan for Energy Efficiency, *National Action Plan for Energy Efficiency Vision for 2025: A Framework for Change*, at ES-1 (2008), *available at* http://www.epa.gov/cleanenergy/documents/vision.pdf.

[34] For more information about the Air Bag and Seat Belt Safety Campaign, visit http://www2.nsc.org/airbag.htm. General Motors and the Insurance Institute for Highway Safety provided the primary funding.

[35] National Safety Council, *National Safety Council Honors Air Bag and Seat Belt Safety Campaign for Lifesaving Achievements* (2006), *available at* http://www.csrwire.com/press/press_release/18157-National-Safety-Council-Honors-Air-Bag-and-Seat-Belt-Safety-Campaign-for-Lifesaving- Achievements.

[36] National Safety Council, Welcome to the Air Bag & Seat Belt Campaign (2004), *available at* http://www2.nsc.org/airbag.htm.

[37] FERC staff defines technical assistance to include, for example, the provision of information, supporting research, funding, and onsite assistance.

[38] Conn. Gen. Stat. § 16a-3a (2008).

[39] For example, see FERC Staff, *2009 Assessment of Demand Response & Advanced Metering,* at 12 (2009) (identifying the steps that states have taken to encourage demand response, such as Pennsylvania, Maryland, New York, New Mexico, Ohio, and Iowa's adoption or enhancement of energy efficiency resource standards that include demand response).

[40] JP Lark, *Michigan's 21st Century Electric Energy Plan* (2007), *available at* http://www.michigan.gov/documents/mpsc/21stcenturyenergyplan_185274_7.pdf.

[41] New Jersey Board of Public Utilities, *New Jersey Energy Master Plan* (2008), *available at* http://www.state.nj.us/emp/docs/pdf/081022_emp.pdf.

[42] 2008 Mass. Acts 50 (amending Mass. Gen. Laws ch. 164B, § 116(a) (2009)).

[43] 2008 Md. Laws 131 (amending MD Code Ann., Pub. Util. Cos. § 7-211(b)(2) (2009)).

[44] 2008 Pa. Laws 129 (amending 66 Pa. Cons. Stat. § 2806.1 (2009)).

[45] S. 39, 67th Gen. Assem., Reg. Sess. (Co. 2009) (amending Colo. Rev. Stat. § 40-9.5-106(2)). ("A cooperative electric association…establishes a graduated rate for increased energy consumption, for energy conservation and energy efficiency purposes, by residential customers that is revenue-neutral for the class…[and] shall give due consideration to the impact of such rates on low-income customers.")

[46] Iowa Code § 476.6.16 (2009). *See* Iowa Utilities Board, *Energy Efficiency in Iowa's Electric and Natural Gas Sectors,* Report to the Iowa General Assembly (2009); Iowa Utilities Board, *Overview of Energy Efficiency Plans and Programs in Iowa* (2009), http://www.state.ia.us/government/com/util/energy/energy_efficiency.html.

[47] "Energy efficiency programs include activities which lessen the amount of heating, cooling, or other forms of work which must be performed, including but not limited to…direct and indirect load control, time-of-use rates…," Iowa Code § 476.6.16(c)(2) (2009) (existing code amended in 2008).

[48] Several states have already implemented demand response programs, from which useful lessons can be drawn. Appendix B discusses specific case studies of demand response programs and policies in various states and regions.

[49] Such a national forum on demand response was jointly conducted by FERC and DOE in February 2002.

[50] These activities would be funded by the Coalition or other private entities rather than the federal government.

[51] Social science research focuses on customer interest in demand response and how customers change their consumption or behavior in response to demand response programs or dynamic pricing.

[52] Pub. L. 111-5, 123 Stat. 115 (2009).

[53] For example, in some localities, the action step for the customer might be, "You need to do nothing," if an opt-out program is implemented that gives the utility, load-serving entity, or other demand response provider the ability to override the end-user's air conditioning controls; for other customers, the action step might be, "Look for price signals, and then manually turn your air conditioner to a higher temperature," if their utilities, load-serving entities, or other demand response providers introduce opt-in dynamic pricing.

[54] Large commercial and industrial customers have greater awareness and knowledge about energy issues; therefore, a mass media campaign may not be appropriate for these customers.

[55] Entities that have already begun compiling data and research from existing demand response programs include the Peak Load Management Alliance and the Demand Response Coordinating Committee, among others.

[56] ENERGY STAR® is a joint voluntary labeling program of EPA and DOE that is designed to identify and promote energy-efficient products to reduce greenhouse gas emissions. For more information about the program, visit http://www.energystar.gov/index.cfm?c=home.index.

[57] For example, the California Statewide Pricing Pilot found that residential, medium commercial and industrial, and large commercial and industrial customers are more responsive to dynamic rates than are small commercial and industrial customers in the absence of automating technologies. See Charles River Associates, Impact Evaluation of the California Statewide Pricing Pilot (2005), *available at* http://sites.energetics.com/madri/ toolbox/pdfs/ pricing/cra_2005_impact_eval_ca_pricing_pilot.pdf.

[58] Note that the technical assistance would help to provide the underlying factual basis for the content of the message.

[59] "Viral marketing" refers to marketing intended to be "spread" by those to whom it is marketed, instead of via traditional media channels like news and advertising. For example, a viral video is produced with the intent that people within the target audience would view it and then share it with their friends and associates, who would in turn share it with their own networks, and so on. This spreading of content through social networks is called viral marketing because of its similarity instructure to the public health model of a virus spreading from its original source.

[60] *DOE, DOE Announces Funding Opportunity to Develop the Smart Grid Information Clearinghouse* (2009), available at http://www.oe.energy.gov/news_room_and_events/ 1223.htm.

[61] Rich Sedano, *Smart Grid Information Clearinghouse*, Presentation to NARUC-FERC Smart Grid Collaborative, available at http://www.narucmeetings.org/Presentations/FercSmartGrid _Sedano.pdf.

[62] *Id.*

[63] *Id.*

[64] *See* California Measurement Advisory Council, http://www.calmac.org/ (last visited Jan. 21, 2010).

[65] *Id.*

[66] *Id.*

[67] See, e.g., California Public Utilities Commission, Decision Adopting Dynamic Pricing Timetable and Rate Design Guidance for Pacific Gas & Electric Company, D. 08-07-045 (2008).

[68] Conn. Gen. Stat. § 16a-3a (2008).

[69] Id., § 16a-3a(c).

[70] *Id.*

[71] The Web-based clearinghouse should serve as a central location for communications materials as well. For information on these materials, see Section 2.2.

[72] DTE Energy, *Interruptible Air Conditioning Rate* (2009), http://www.dteenergy.com/ residentialCustomers/billingPayment/rates/electric/interruptibleRates.html.

[73] For more information about the Flex Your Power campaign, visit http://www.fypower.org/.

[74] Assessment of Demand Response and Advanced Metering, *supra* note 18.

[75] NERC, *Data Collection for Demand-Side Management for Quantifying Its Influence on Reliability: Results and Recommendations* (2007).

[76] For more detailed information on dynamic rate design see, for example, Faruqui A, Hledik R, "Transition to Dynamic Pricing," *Public Utilities Fortnightly* (March 2009), and Faruqui A et al., *Rethinking Rate Design Demand Response* (2007) (issue paper prepared for the Demand Response Research Center, Lawrence Berkeley National Laboratory).

[77] More information and a summary of recent activity can be found at http://www.nist.gov/ smartgrid/.

[78] *See id.*

[79] *See id.*

[80] *See* Lawrence Berkeley National Laboratory, *Open Automated Demand Response Communication Specification* (Version 1.0), (2009), *available at* http://drrc.lbl.gov/openadr/ pdf/cec-500-2009-0.63.pdf (prepared for the California Energy Commission, Public Energy

Research Program). This report "outlines a communications exchange signals specification to enable demand response in end- use participant or customer systems."

[81] For instance, SAE International has released a standard that provides a standard interface between plug-in hybrid and battery-electric vehicles and electrical charging systems. *See* SAE International, *New SAE International Standard Defines Electric-Vehicle Charging Coupler* (2010), *available at* http://www.sae.org/servlets/pressRoom?OBJECT_TYPE =PressReleases&PAGE=showRelease&RELEASE _ID=1141.

[82] NAESB, *M&V for Demand Response Programs* (2009), *available at* http://naesb.org/pdf4/ dsmee_group2_070709reqcom_attach.doc.

[83] The model is available at http://www.ferc.gov/industries/electric/indus-act/demand-response/ dr- potential.asp.

[84] For more information about the InterAct tool, visit http://www.pge.com/mybusiness/ energysavingsrebates/demandresponse/tools/.

[85] DOE developed this directory because its Office of Building Technology, State and Community Program develops software tools to help researchers, designers, architects, engineers, builders, code officials, and others involved in the building life cycle to evaluate and rank potential energy efficiency technologies and renewable energy strategies in new or existing buildings. To learn more about the directory and tools, visit http://apps1.eere. energy.gov/buildings/tools_directory/.

[86] To learn more about the Demand Response Quick Assessment Tool, visit http://apps1. eere.energy.gov/buildings/tools_directory/software.cfm/ID=522/pagename=alpha_li

[87] For more information about RMI's dispatch model, visit http://ert.rmi.org/research/next-generation- utility.html.

[88] CPUC, *California Standard Practice Manual: Economic Analysis of Demand-Side Programs and Projects* (2002), *available at* http://drrc.lbl.gov/pubs/CA-SPManual-7-02.pdf.

[89] *Preventing Undue Discrimination and Preference in Transmission Service*, Order No. 890, FERC Stats. & Regs. ¶ 31,241, at P 479, *order on reh'g*, Order No. 890-A, FERC Stats. & Regs. ¶ 31,261 (2007), *order on reh'g*, Order No. 890-B, 123 FERC ¶ 61,299 (2008), *order* on reh'g, Order No. 890-C, 126 FERC ¶ 61,228 (2009), Order No. 890-D, 129 FERC ¶ 61,126 (2009).

[90] Nexus Energy Software et al., *Information Display Pilot* (2005).

[91] The Brattle Group, *BGE's Smart Energy Pricing Pilot: Summer 2008 Impact Evaluation*, (2009) (prepared for Baltimore Gas & Electric Company).

[92] Connecticut Light and Power, *Results of CL&P Plan-it Wise Energy Pilot,* Docket No. 05-10-03RE01, Compliance Order No. 4 (2009).

[93] Social science research focuses on customer interest in demand response and how customers change their consumption or behavior in response to demand response programs or dynamic pricing.

[94] More information about these programs can be found at http://www.srpnet.com/prices/ home/tou.aspx (Salt River Project) and http://www.aps.com/aps_services/residential/ rateplans/Default.html (Arizona Public Service Company).

[95] eMeter, *Demand Response: An eMeter White Paper Presentation* (2008).

[96] Salt River Project, *Navigating Change: SRP 2009 Annual Report,* at 7 (2009), *available at* http://www.srpnet.com/about/financial/2009AnnualReport/siteassets/pdfx/2009SRPAnnual Report.pdf.

[97] *Id.*

[98] Cal. Code Regs. tit. 20, §§ 1601–1608 (2009) (efficiency standards for appliances); Cal. Code Regs. tit. 24.

[99] California Energy Commission, *2007 Integrated Energy Policy Report,* at 3–4 (2007).

[100] State of California, *Energy Action Plan,* at 4 (2003).

[101] *Id.*

[102] *See, e.g.,* Pacific Gas & Electric Company, *Decision Adopting Dynamic Pricing Timetable and Rate Design Guidance for Pacific Gas & Electric Company,* D. 08-07-045 (2008).

[103] For more information about the Flex Your Power campaign, visit http://www.fypower.org/.

[104] For more general information about the pilot program and its results, see Charles River Association, *Impact Evaluation of the California Statewide Pricing Pilot* (2005), *available at* http://sites.energetics.com/madri/toolbox/pdfs/pricing/cra_2005_impact_eval_ca_pricing_pilot.pdf.

[105] *See id.*

[106] San Francisco Community Power, *Statewide Pricing Pilot — Track B End of Pilot Focus Group Results 2* (2005), *available at* http://www.sfpower.org/SFCP_TrackB_FocusGroups.pdf (finding strong statistical support that customers wanted to remain on the critical peak pricing rates).

[107] Conn. Gen. Stat. § 16a-3a (2008).

[108] *Id.*, § 16a-3a(c).

[109] *Id.*

[110] Connecticut Light and Power, *Plan-It Wise Energy Program,* http://www.cl-p.com/home/saveenergy/goinggreen/planitwise.aspx (2009).

[111] "CL&P Energy Program Pilot Customers Respond to Price Signals," *Transmission & Distribution World* (2010), *available at* http://tdworld.com/customer_service/clp-energy-program-20100101/.

[112] Del. Code Ann. tit. 26, § 1002(4) (2009).

[113] *Id.*

[114] Gulf Power, *Energy Select* (2009), http://www.gulfpower.com/energyselect/index.asp.

[115] Gulf Management Company, *Good Cents SELECT: Advanced Energy Management Program*, at slide 6 (2006), *available at* http://www.ewh.ieee.org/r3/nwflorida/presentations/01_19_06.ppt.

[116] *Id.*

[117] Thompson J, Gulf Power, *Presentation for the Southeast Energy Efficiency Meeting,* at 14 (2007), *available at* http://www.epa.gov/RDEE/documents/southeast- meeting/thompson_southeast_energy_efficiency_meeting_atlanta_ga.pdf (discussing real-time pricing and the Good Sense *SELECT* program at Gulf Power).

[118] E-mail from Julie Kreafle, Manager, Product Marketing, Florida Power and Light Company, to Ahmad Faruqui (Sep. 16, 2009, 9:47 PM ET) (on file with The Brattle Group).

[119] O'Sheasy M, "How to Buy High and Sell Low: Spot Priced Electricity Offers Financial Rewards," *Pricing in Competitive Electricity Markets* 295 (2000).

[120] Braithwait S, *Evaluating Changes in RTP Price Response — The Effect of Market and Environmental Conditions*, prepared for the International Energy Program Evaluation Conference Proceedings in New York (2005).

[121] *Id.*

[122] 2005 Ill. Laws 977 (amending 220 Ill. Comp. Stat. Ann. 5/16-101A, 16-102, 16-107).

[123] Comverge, Inc., *ComEd Residential Real-Time Program – 2007 Annual Report,* Illinois Commerce Commission Docket No. 06-0617 (filed on April 30, 2008), *available at* http://www.icc.illinois.gov/docket/files.aspx?no=06-0617&docId=123573.

[124] Ameren Illinois Utilities, *RSP Rider — Power Smart Pricing — 2007 Annual Report,* Illinois Commerce Commission Docket No. 06-0691 (filed April 15, 2008), *available at* http://www.icc.illinois.gov/docket/files.aspx?no=06-0691&docId=123030.

[125] Comverge, Inc., *ComEd Residential Real-Time Pricing Program – 2008 Revised Annual Report,* Docket No. 06-0617 (filed on April 28, 2009), *available at* http://www.icc.illinois.gov/docket/files.aspx?no=06-0617&docid=136091.

[126] *Ameren Illinois Utilities, Rider PSP — Power Smart Pricing — 2008 Annual Report,* Illinois Commerce Commission Docket No. 06-0621 (filed on April 30, 2009), *available at* http://www.icc.illinois.gov/docket/files.aspx?no=06-0691&docId=136151.

[127] 2008 Mass. Acts 50 (amending Mass. Gen. Laws ch. 164B, § 116(a) (2009)).

[128] Governor Deval Patrick, *Governor Patrick Signs Energy Bill Promoting Cost Savings, Renewable and Clean Energy Technology* (2008), *available at* http://www.mass.gov/?

pageID=gov3pressrelease&L=1&L0=Home&sid=Agov3&b=pressrelease&f=080702_bill_energy_clean&csid=Agov3.

[129] Lark JP, *Michigan's 21st Century Electric Energy Plan* (2007), *available at* http://www.michigan.gov/documents/mpsc/21stcenturyenergyplan_185274_7.pdf.

[130] Consumers Energy Company, *Electric Generation Alternatives Analysis for Proposed Permit to Install No. 341-07 for an Advanced Supercritical Pulverized Coal Boiler at the Karn-Weadock Generating Station,* MPSC Case No. U-15996 (filed with the Mich. Pub. Serv. Commn. June 5, 2009).

[131] In 2018, air conditioning load management is projected to be able to reduce peak demand by 308 MW. Customer demand response programs associated with AMI are projected to provide further peak load reduction of 212 MW by 2018.

[132] E-mail from Mark Lesiw, AMI Senior Program Manager of Load Management, Consumers Energy, to Ahmad Faruqui (Sep. 2, 2009, 1:19 PM ET) (on file with The Brattle Group).

[133] *Id.*

[134] DTE Energy, *Interruptible Air Conditioning Rate* (2009), http://www.dteenergy.com/residentialCustomers/billingPayment/rates/electric/ interruptibleRates.html.

[135] E-mail from Mark Williamson, Manager, Load Research, DTE Energy, to Ahmad Faruqui (Aug. 31, 2009, 1:21 PM ET) (on file with The Brattle Group).

[136] *Id.*

[137] E-mail from Mark Williamson, Manager, Load Research, DTE Energy, to Ahmad Faruqui (Aug. 31, 2009, 9:48 AM ET) (on file with The Brattle Group).

[138] *See* MacDonald T, *The Localized Substation Load Relief Program at Detroit Edison*, Presentation for AEIC National Conference (2006), PowerPoint *available at* http://www.aeic.org/load_research/docs/15_2006_09_The_Localized_Substation_Load_Relief_Progra m_at_Detroit_Edison_presentation_.ppt.

[139] *See id.*

[140] For more information about this pilot program, please visit http://www.powercentsdc.org/.

[141] E-mails from Chris King, eMeter, to Ahmad Faruqui (Aug. 30, 2009, 2:08 AM ET; Sept. 1, 2009, 10:15 AM ET) (on file with The Brattle Group).

[142] The information is adapted from the ISO/RTO Council's North American Wholesale Electricity Demand Response Program Comparison, *available at* http://www.isorto.org/atf/cf/%7B5B4E85C6-7EAC-40A0-8DC3-003829518EBD%7D/IRC%20DR%20M&V%20Standards%20Implementation%20Comparison%20(20 09-04-28).xls.

In: Demand Response ...
Editor: Adam R. Peterson ISBN: 978-1-62948-023-7

Chapter 2

IMPLEMENTATION PROPOSAL FOR THE NATIONAL ACTION PLAN ON DEMAND RESPONSE*

U.S. Department of Energy,
Federal Energy Regulatory Commission

EXECUTIVE SUMMARY

The staff of the Federal Energy Regulatory Commission (FERC) and the U.S. Department of Energy (DOE) developed this implementation proposal as required by section 529 of the Energy Independence and Security Act of 2007 (EISA).[1] In particular, this proposal complies with EISA's mandate "to submit to Congress a proposal to implement the [National] Action Plan [on Demand Response], including specific proposed assignments of responsibility, proposed budget amounts, and any agreements secured for participation from State and other participants."[2]

The objective of the proposal is to implement the National Action Plan to achieve the nation's demand response potential. FERC staff and DOE believe that every element of the National Action Plan serves an important part in achieving this potential. This proposal recognizes, however, that it will take public and private actors to effectuate the Plan and, as a result, not

---

* This Report is prepared by staff of the Federal Energy Regulatory Commission and the U.S. Department of Energy on July 2011.

every element of that plan may be implemented. As such, this implementation proposal identifies areas where FERC staff and DOE can leverage existing initiatives related to demand response to implement the National Action Plan. At the same time, it recognizes and encourages non-federal government entities to become involved in implementing the National Action Plan through participation in a coalition, or any private or non-federal governmental organizations.[3] FERC staff's primary role with a coalition, or any private or non-federal governmental organizations, will be to monitor, consult with, and provide recommendations regarding the implementation of the National Action Plan. A coalition, or any private or non-federal governmental organizations, should consider the viewpoints of a broad range of demand response stakeholders and private sector organizations. Action under this proposal will not require any new federal appropriations.

The following chart summarizes the appropriate actor identified for each activity, as well as any actions that can be leveraged to support the accomplishment of each activity. As explained above, the activities discussed in this proposal were introduced and recommended in the National Action Plan to achieve the objectives laid out in EISA. As such, the activity headings track the headings from the National Action Plan.

Activity	Appropriate Main Actor	Supporting Actions
TECHNICAL ASSISTANCE TO STATES		
Establish a National Forum	DOE, FERC and a coalition, or any private or non-federal governmental organizations	
Conduct Informational and Educational Sessions for Policymakers and Regulators	DOE	• DOE support through the Mid-Atlantic Distributed Resource Initiative, the Midwest Demand Response Initiative, and the Pacific Northwest Demand Response Project (PNDRP). • DOE grants to National Association of Regulatory Utility Commissioners (NARUC), the National Conference of State Legislatures, the National Governors Association, and the National Association of State Energy Officials.

Activity	Appropriate Main Actor	Supporting Actions
		• FERC and DOE support through the FERC-NARUC Smart Response Collaborative. • A coalition, or any private or non-federal governmental organizations, may sponsor additional educational sessions.
Sponsor Technical Papers	A coalition, or any private or non-federal governmental organizations	• DOE commitments to fund technical studies of: 1) customer response to and acceptance of dynamic pricing funded through the Smart Grid Investment Grant (SGIG) program, 2) customer behavior in response to energy use feedback and information funded through the SGIG program, and 3) the role of demand response in transmission system planning, and other topics as needed.
Technical Assistance to States for Demand Response	DOE	• DOE provision of technical assistance to states that wish to support the development of demand response.
Build a Panel of Demand Response Experts	A coalition, or any private or non-federal governmental organizations	
Establish a Demand Response Grant Program	A coalition, or any private or non-federal governmental organizations	• DOE commitment of $4 billion of Recovery Act funds to 131 entities through the SGIG and Smart Grid Demonstration programs.
National Communications Program		
Develop a Communications Umbrella	A coalition, or any private or non-federal governmental organizations	• DOE will compile customer education efforts funded through the SGIG program. • DOE will compile results from consumer behavior studies focused on price-responsive demand funded through the SGIG program. • DOE and the Environmental Protection Agency (EPA)

Activity	Appropriate Main Actor	Supporting Actions
		share information developed through the State Energy Efficiency Action Network's "Customer Information and Behavior Working Group."
Local Implementation	A coalition, or any private or non-federal governmental organizations/demand response providers/ load-serving entities	
Direct Outreach to States, Policymakers, and Partners	A coalition, or any private or non-federal governmental organizations	
Tools and Materials		
Develop a Web-Based Clearinghouse of Demand Response Materials	DOE	• DOE will examine the feasibility of integrating a web-based demand response clearinghouse with the DOE-sponsored Smart Grid Information Clearinghouse.
Develop or Enhance Demand Response Estimation Tools and Methods	A coalition, or any private or non-federal governmental organizations	• DOE is assisting the development of a research and data analysis plan for SGIG project data. • DOE has established guidance on the identification of SGIG program metrics and impacts. • Entities funded by DOE through the Recovery Act to conduct interconnection-wide transmission planning are examining demand-side assumptions and impacts as part of their analysis and planning efforts. • Entities funded by DOE through the Recovery Act to conduct interconnection-wide transmission planning will also develop tools and methods as needed to better incorporate demand side measures into their transmission plans. • DOE is funding

Activity	Appropriate Main Actor	Supporting Actions
		development of a cost-effectiveness screening tool for PNDRP. • DOE will work with the state regulatory community to assess the extent to which federally-sponsored research on cost-effectiveness guidelines for demand response resources would be useful to state PUCs in various regions. • The National Institute of Standards and Technology (NIST) and the North American Energy Standards Board (NAESB) are developing standards for the provision of energy price information. • DOE's ARPA-e program is funding consumer behavior and energy use research.

I. INTRODUCTION

A. Legislative Mandate

As set forth in section 529 of EISA, Congress required FERC to study and develop a process to achieve the Nation's demand response potential. The statute required three reports. First, it required FERC to prepare a National Assessment of Demand Response Potential (National Assessment).[4] The National Assessment not only evaluated the nationwide potential for demand response over five- and ten-year horizons, it made policy recommendations to achieve the Nation's estimated potential. The National Assessment also identified barriers that have limited the deployment of demand response and recommended policies to overcome these barriers. FERC staff submitted the National Assessment to Congress on June 17, 2009.[5]

Second, EISA required the FERC to develop a National Action Plan.6 This report needed to meet three objectives:

1. Identification of the requirements for technical assistance to States to allow them to maximize the amount of demand response resources that can be developed and deployed.
2. Design and identification of requirements for implementation of a national communications program that includes broad-based customer education and support.
3. Development or identification of analytical tools, information, model regulatory provisions, model contracts, and other support materials for use by customers, States, utilities and demand response providers.[7]

FERC staff publicly released the National Action Plan on June 17, 2010.[8]

Finally, EISA requires the FERC, together with the Secretary of Energy, to submit to Congress a proposal to implement the National Action Plan. This implementation proposal must include "specific proposed assignments of responsibility, proposed budget amounts, and any agreements secured for participation from state and other participants."[9]

B. Approach to Developing the National Action Plan and Implementation Proposal

Per Congress' direction, the National Action Plan identifies means for the states to maximize the amount of cost-effective demand response achieved in the nation; design requirements for a communications program and materials to support demand response; and analytical tools for demand response design, evaluation, and analysis. The Plan was developed with input and participation from a broad range of stakeholders, including local, state, and federal government officials; customer representatives; industry representatives; and nongovernmental groups. As such, it recognizes the important role that these stakeholders have in the successful development of demand response and the implementation of the National Action Plan, particularly state and local governments.

This implementation proposal builds on the National Action Plan by identifying actions that can be taken by federal, state, and private sector organizations to implement the Plan.[10]

However, a primary focus of this implementation proposal is the activities that are appropriately undertaken by FERC staff and DOE to implement the National Action Plan, including interaction and coordination with private sector and fellow government partners to advance demand response activities through a coalition, or any private or non-federal governmental organizations. This implementation proposal also provides the views of FERC staff and DOE on some of the National Action Plan activities that are appropriate for private sector and state action, and offers some suggestions for how non-federal entities might choose to implement these activities. The states and private sector entities will decide whether and how to pursue these recommendations and other activities relevant to demand response, depending on their particular circumstance.

As required by EISA, FERC staff and DOE propose assignments of responsibility for implementing the National Action Plan. The responsibility for implementing many of the activities has been assigned to the private sector, ideally through a broad coalition of demand response stakeholders, or any private or non-federal governmental organizations that coordinate and cooperate to implement the Action Plan. However, the implementation proposal also recognizes that there are ongoing demand response programs and efforts, some of which are being funded by federal government agencies. Since passage of EISA in December 2007, there is a renewed awareness of demand response and the different roles it can play. Congress passed the American Recovery and Reinvestment Act of 2009 (Recovery Act) under which DOE has awarded $4 billion to stimulate development of the smart grid, including a substantial number of projects deploying some type of demand response. Given this and other activities, FERC staff and DOE believe that existing federal resources and programs can be harnessed to achieve many of the elements of the Plan. Accordingly, this implementation proposal relies as much as possible on existing public programs, and encourages new private involvement through a coalition, or any private or non-federal governmental organizations, to implement the activities outlined in the National Action Plan and, in turn, achieve the objectives of EISA.

This implementation proposal contemplates that, to the extent permitted by federal law and regulations, FERC staff and DOE may support certain activities included in the National Action Plan to the degree that those activities entail research, tool development, and information-sharing among the community of demand response stakeholders, and provide resources and technical assistance that help others implement these objectives.

DOE, and in some cases FERC, can meet with any coalition, or any private or non-federal governmental organizations, and share information and coordinate activities with any of these entities, as allowed to under Federal law. DOE, and in some cases FERC, can also develop reports and resources individually, in concert with, or on behalf of the National Action Plan, and consider policy and action recommendations from a coalition, or any private or non-federal governmental organizations. However, it would not be appropriate for DOE and FERC to join a coalition, provide general operational funds for a coalition, take direction from a coalition, or any private or non-federal governmental organizations, or give direction to others such as states through a coalition, or any private or non-federal governmental organizations.

II. Implementation Proposal

A. Coalition

As explained in the National Action Plan, a broad range of stakeholders must be involved in the Plan's implementation in order for it to succeed. Therefore, the National Action Plan calls for a broad, inclusive coalition to ensure that the National Action Plan reflects regional differences and taps into the expertise of stakeholders. To be effective, a coalition, or any private or nonfederal governmental organizations, should consider the efforts and viewpoints of state/local government officials, utilities/load-serving entities, demand response providers, regional transmission organizations/independent system operators, consumer advocates, commercial/industrial customers, the federal government, existing coalitions, and other stakeholders.[11]

B. Technical Assistance to States

1. Establish a National Forum

The National Action Plan calls for a national forum on demand response to be held for state public utility commissions, state energy offices, gubernatorial offices, state legislators, state consumer counsels, utilities, demand response providers, other key stakeholders, interested consumers, and federal agencies. Such a stand-alone, one-day session will provide a useful opportunity to publicly present the overall vision of the National Action Plan, and to share ideas, examine barriers, and explore solutions. Given current financial and time burdens upon governments and businesses alike, this national forum event should be offered with both in-person and webcast access, to maximize participation by a diverse community of stakeholders at minimal cost.

FERC staff and DOE can and should sponsor this forum in 2011, as they did in 2002 with their first national forum on demand response. Non-federal members of the demand response community could use the forum as an opportunity to convene stakeholders to discuss the organizational structure and leadership of a coalition, or any private or non-federal governmental organizations used to forward the work of the National Action Plan.

2. Conduct Informational and Educational Sessions for Policymakers and Regulators

DOE and FERC engage in a number of activities that provide informational and educational support that advance demand response. At the request of regional organizations, DOE has provided facilitation support to conversations by state electricity regulators and others through the New England Demand Response Initiative (2001-2006), the Mid-Atlantic Distributed Resources Initiative (2004-present), the Midwest Demand Response Initiative (2007-present), and the Pacific Northwest Demand Response Project (2007-present). Every one of these regional fora has been providing informational and educational sessions on demand response to the state utility commissions that sponsor them. Similarly, FERC, in collaboration with state public utility commissioners, and with support from DOE on specific activities, has been sponsoring informational and educational sessions on demand response for policymakers and regulators through the thrice-yearly FERC-National Association of Regulatory Utility Commissioners (NARUC) Demand Response Collaborative (that began in 2007 and has recently joined with the FERC-NARUC Smart Grid Collaborative to become the FERC-NARUC Smart Response Collaborative). DOE has funded additional education and information sessions for regulators and state officials through past and current grants to NARUC, the National Conference of State Legislatures, the National Governors Association, the National Association of State Energy Officials, and the Western Governors' Association, supporting both training costs and travel costs for state attendees.

Over the past 10 years, DOE has funded and provided technical assistance on demand response and smart grid issues to approximately 30 states. This assistance is provided upon request and is not intended to influence decision making by state and local officials. These presentations and consultations entailed both direct education and training on general demand response issues and targeted consultation and technical assistance on selected issues, including conveying the results of DOE-funded technical research and dissemination of technical papers by DOE and others.

Going forward, DOE plans to continue offering technical assistance to state regulators on demand response. In recent years, requests for technical assistance on demand response issues have often been in the form of requests for technical assistance on smart grid implementation issues. Thus, DOE now offers technical assistance to state public utility commissions on demand response and smart grid, either when those requests arise from a specific demand response regulatory proceeding/workshop/issue or as part of a broader set of implementation issues that confront policymakers and regulators on smart grid.

At the request of NARUC, DOE plans to help NARUC develop additional emphasis on demand response educational and informational sessions at regional meetings of state regulators; for example, NARUC, using DOE funds, organized a "smart grid school" for the June 2010 annual meeting of the Mid-Atlantic Conference of Regulatory Utility Commissioners.

3. Sponsor Technical Papers

The National Action Plan recommends development of numerous technical papers to generate new knowledge and synthesize existing knowledge to inform state officials about demand response program design, products, technologies, existing state incentives, markets, and existing state legislation. A coalition, or any private or non-federal governmental organizations, could oversee development of these papers, which would be targeted to state officials.

In addition, DOE is committed to sponsoring research projects and studies that will lead to technical reports on selected, high-priority demand response topics. For example, DOE has made specific commitments to fund technical studies in the following areas:

Customer Response to and Acceptance of Dynamic Pricing

- Several of the Recovery Act-funded smart grid grants will implement consumer behavior studies to examine the influence of information technology, dynamic prices, and education on consumer behavior. DOE is working closely with each of these funded recipients for them to design and implement their own two- to three-year studies. These studies will result in reports, including a meta-analysis by DOE, as well as interval meter data annotated at an appropriate privacy level with demographic information.

Customer Behavior in Response to Energy Use Feedback and Information

- DOE's Advanced Research Projects Agency-Energy (ARPA-E) energy research and development innovations program has issued a $5 million grant to Stanford University's Precourt Institute to study customer behavior with respect to energy efficiency that will have implications for demand response.
- DOE and the Environmental Protection Agency (EPA) support the State Energy Efficiency Action Network's "Customer Information and Behavior Working Group," which is studying consumer behavior with respect to energy efficiency and could generate pertinent information.

Role of Demand Response in Transmission System Planning

- At the request of the five entities performing Recovery Act-funded interconnection-wide transmission planning and analysis work, DOE is funding work by two national laboratories (Lawrence Berkeley National Laboratory and Oak Ridge National Laboratory) to provide technical assistance to the interconnection-wide transmission planning and analysis organizations[12] on demand-side resource assessment, methods to incorporate some dynamic pricing into long-term load and resource scenarios, and data development and tools needed to better include the demand side in transmission planning. In addition, the five interconnection-wide entities are including various demand-side scenarios that include demand response in their transmission planning analyses.

The current, ongoing research efforts, as well as past research and studies, represent a significant investment by DOE in improving knowledge of key demand response impact and use issues that are appropriate for federal action because of their national reach.

Going forward, FERC staff and DOE will continue to talk with the demand response community and monitor Smart Grid Investment Grant (SGIG) project developments to identify those opportunities where there is little private or non-federal action *and* where federal expertise, quickly applied, can add value and insight.

The National Action Plan also identified a number of topics that deserve additional research. While FERC staff and DOE believe that it is appropriate for the federal government to conduct research on some of those topics, Appendix A of this document includes a list of other research studies that, in some cases, may be more appropriately performed by individual technology vendors, demand response providers, and utilities that will directly benefit from the research.

4. Technical Assistance to States for Demand Response

The National Action Plan recommends providing technical assistance for states, including direct assistance to states to implement demand response activities, supporting the existing FERCNARUC Smart Response Collaborative, and providing access to federal demand response experts.

As stated above, DOE has a long history of, and currently provides, technical assistance to states on demand response. DOE intends to continue doing so, together with necessary research to better inform and support such assistance.[13] The scope of this work has addressed the development of demand response programs, technologies, services infrastructure, and dynamic pricing tariffs.

Through its annual advanced metering and demand response reports, FERC staff provides up-to-date information to states and other interested parties on the penetration of advanced meters, demand response programs, and trends in demand response policy. FERC staff's National Assessment provided state-by-state estimates of demand response potential and an analysis tool that states can use to develop their potential estimates.

In this implementation proposal, FERC staff and DOE believe that it is appropriate to link informational and educational sessions for policymakers and regulators (see section II.B.) as part of the overall technical assistance activity for demand response which often involves direct support to state, regional, and national organizations on selected demand response policy, program design, and/or technology issues.

5. Build a Panel of Demand Response Experts

The National Action Plan recommended creating a panel of demand response experts to: 1) inform various stakeholders about demand response programs, products, technologies, and incentives and their benefits; 2) serve as a resource, upon request, for state and local government officials as they consider demand response programs and policies; and 3) serve as speakers at events.

This past decade has seen a proliferation of privately- and publicly-sponsored conferences, meetings, seminars, classes and webinars on demand response. The hundreds of individuals from diverse backgrounds and institutions who have participated on those agendas, as well as the number of published papers, blog entries, and press releases on demand response, make it clear that there is neither lack of expertise nor a lack of those willing to share their views on demand response.

Moreover, both DOE and FERC have hired various researchers and consultants to conduct or assist with formal demand response analytical and training efforts. For example, FERC has brought in such expert assistance for its annual Demand Response and Advanced Metering reports, Demand Response Potential Assessment, and the National Action Plan. DOE has funded a suite of activities through technical assistance to states and research published by the Lawrence Berkeley National Laboratory, DOE Office of Electricity Delivery and Energy Reliability's Load as a Resource research and development program, the Regulatory Assistance Project, NARUC, the National Governors Association, the National Conference of State Legislatures, and other entities.

As a result, there are many established experts, and FERC staff and DOE do not believe it is necessary or appropriate for federal agencies to identify or pre-qualify private sector experts. Private or non-federal governmental organizations, or a coalition, may choose to maintain rosters of demand response experts, develop procedures to identify "experts," and may choose to finance their activities on behalf of their members and constituents.

6. Establish a Demand Response Grant Program

The National Action Plan recommends creation of a grant program to fund demand response activities. Under the Recovery Act, DOE has committed $4 billion of federal funds in the Smart Grid Investment Grant and Smart Grid Demonstration programs to 131 organizations for smart grid projects. Many of these projects directly support the goals of the National Action Plan (i.e., to accelerate the understanding and achievement of demand response across the nation) and involve deployment of enabling technologies and customer information/feedback systems to facilitate demand response. Smart grid investment grants and demonstration projects will help fund the installation of 15.5 million new smart meters, 6500 distribution circuits with automated equipment, and 800 networked phasor measurement units at 131 utilities and other grantee locations.

Many of the smart meter installations will include both information and control technologies such as in-home displays and programmable, communicating thermostats with corresponding demand response programs. A number of consumer behavior studies will be performed on those using the installed equipment in order to examine factors that influence the participation of consumers in dynamic pricing programs, as well as the influence of these programs and enabling technologies on customer response.

Given this unprecedented infusion of federal funds to support smart grid and demand response, FERC staff and DOE believe that additional federal funding is not required for a demand response grant program to support further pilot and/or demonstration projects. However, nonfederal entities acting through a coalition, or any private or non-federal governmental organizations, may determine that additional non-federal sector funding for particular pilot and/or demonstration projects, or assessment of demand response technologies would be helpful and may award such funds, whether independently or in conjunction with a coalition, or any private or non-federal governmental organizations.

C. National Communications Program

FERC staff and DOE propose that a coalition, or any private or non-federal governmental organizations, execute the communications program recommended in the National Action Plan. Given the experience, expertise, and interest of the private sector in the area of demand response customer communications, the private sector should assume the responsibility of creating, organizing, and administering the national communications program.

However, FERC staff and DOE propose that the federal government provide limited support to the communications program in specific research areas, which are listed here, and as well discussed further below:

- DOE will compile customer education efforts funded through its Smart Grid Investment Grant program.
- DOE will compile results from consumer behavior studies focused on price-responsive demand funded through the SGIG program.
- DOE and the EPA will share information developed through the State Energy Efficiency Action Network's "Customer Information and Behavior Working Group."

Though non-federal organizations will be responsible for implementation of the majority of national communication program activities, the activities outlined in the National Action Plan provide a helpful roadmap of how to achieve the EISA's requirements.

1. Develop a Communications Umbrella

The Communications Umbrella would create and provide a consistent, yet flexible, research-based message framework to provide communications tools, support, and advice that is adaptable for use by interested local demand response implementers. Because large commercial and industrial customers tend to be more sophisticated about energy issues, the Communications Umbrella would also develop specific communications support materials for these customers.

While the vast majority of the Communications Umbrella will need to be funded by private sources, the Federal government can provide limited support for the Communications Umbrella through focused technical research that is in the public interest. For example, several of the Recovery Act-funded smart grid grants will implement consumer behavior studies to examine the influence of information technology, dynamic pricing and education on consumer behavior. DOE is working closely with each of these funded recipients as they design and implement their own two- to three-year studies. These studies will result in reports, including a meta-analysis by DOE. DOE will also report on other Recovery Act-funded smart grid grants that include consumer education, including any lessons learned these grantees may offer on their experiences.

The information and insights gained from these studies should assist coalition or any private or non-federal governmental organization efforts, particularly in the areas of customer response to dynamic pricing and enabling technologies. However, most of the information and analysis from these Smart Grid Investment Grants will not be available until 2012-2013. A coalition, or any private or non-federal governmental organizations, in partnership with utilities and other service providers, should build on DOE's work to further refine and understand the effectiveness of various messages and communications strategies.

a. Program Development

During development of the communications program, a coalition, or any private or non-federal governmental organizations may wish to study how local entities have communicated about demand response implementation across the country and internationally, including messages, focus group feedback, advertisements, and other components. Steps include compiling case studies, identifying demonstration program sites, and conducting program evaluation. Compiling and studying demand response implementation information would support local demand response implementers in their individual efforts and inform the foundational customer research performed during message framework development. Program development could include the selection of demonstration program sites from among those states, local load-serving entities, utilities, and other demand response providers that have requested support and offered to share information about their program's communications and marketing. DOE and EPA could assist by sharing information about work done on customer behavior and interest in demand-side messages (e.g., ENERGY STAR®). Where appropriate, the compilation of this information could be posted on the web-based demand response clearinghouse recommended in FERC's Action Plan.

As part of program development, the National Action Plan recommends the compilation of data. This activity involves collecting data on demand response program implementation pertaining to, among other items, message testing and customer behavior. DOE proposes to post demand response communications case studies in the Smart Grid Information Clearinghouse. DOE encourages a coalition, or any private or non-federal governmental organizations, and SGIG project managers to coordinate their case study templates so that demand response messages and materials impacts are integrated into broader project evaluations.

b Message Framework Development

The National Action Plan emphasizes the importance of message framework development, creating a broad message framework and outreach strategies that can be customized by local stakeholders. Steps include conducting foundational market research, developing adaptable consumer-friendly messages, positioning demand response as a part of smart energy use, and developing communications materials, tactics, toolkits, and support materials tailored to large commercial and industrial customers. By doing so, the national communications program would build upon the lessons learned from other related communications efforts. The Communications Umbrella message framework should yield more effective communications about demand response to customers.

Work being sponsored on customer behavior by DOE and EPA could also provide input to coalition, or other private or non-federal governmental organizations, communications efforts. A work group of the State Energy

Efficiency Action Network effort, sponsored by DOE and EPA, is examining how information impacts customer behavior with respect to energy efficiency. Insights from this work may be useful and supportive of focused coalition or other non-federal governmental or private research on demand response message research.

The Communications Umbrella would involve reframing the concept of demand response with more customer-friendly terminology to better explain the concepts of energy use reduction, load shifting, and managing energy usage in response to price or non-price signals. Under the Communications Umbrella, the coalition, or other private or non-federal governmental organizations, should develop a message framework with persuasive, adaptable messages aimed at various audience segments, all of which could be tailored by interested local stakeholders.

The Communications Umbrella calls for developing and providing communications materials based on the previously described research that local demand response providers can use to create or support local program communications. Additional materials would be tailored to large commercial and industrial customers and would inform these customers of opportunities to provide demand response and the possible benefits of participation.

2. Local Implementation

To meet EISA's requirement for broad-based customer education and support, the National Action Plan calls for a way to advise and support states and localities— governments, utilities, and demand response providers—on communications best practices as they adapt the Communications Umbrella to increase local users' demand response participation in demonstration programs. This element includes the provision of logistical support and coordination for communications and local demand response implementation.

The National Action Plan lists elements of support including provision of the initial Communications Umbrella materials and messages and use of diverse public relations, advertising, viral and grass roots marketing, all linked to a national communications program. Support materials should be designed to be "plug and play" so that local entities can either use all available messages and materials or choose which elements to use.

3. Direct Outreach to States, Policymakers and Partners

The National Action Plan-recommended activities include communications assistance to states and policymakers using meetings, workshops, message guides, and potentially national spokespersons and public educational efforts on demand response. This effort could include assisting and informing state and local governing officials and policymakers about demand response in areas where demand response programs, products, technologies, and initiatives are not yet available. A coalition, or any private or non-federal governmental organizations, may provide assistance when requested, and through regional workshops.

The National Action Plan also calls for the development of partnerships among a coalition, or any private or non-federal governmental organizations, and corporations and organizations that represent various customer classes, trade associations, and other groups. Partnerships would be formed based upon whether organizations can reach key audience segments, are relevant to advancing the issue, and have the ability to commit resources to support the communications effort. Effective partnerships provide another avenue for effective demand response communications.

D. Tools and Materials for Use by Customers, States and Demand Response Providers

1. Develop a Web-Based Clearinghouse of Demand Response Materials

Currently, much of the available information on demand response is being collected by several different entities with little coordination. The National Action Plan calls for the development of a web-based clearinghouse that will facilitate the collection and delivery of up-to-date information and analyses on demand response. Given the overlap between demand response and smart grid, DOE will examine the feasibility of integrating the Action Plan's web-based demand response clearinghouse with the DOE-sponsored Smart Grid Information Clearinghouse.[14] In addition to DOE's efforts, a coalition or other private or non-federal governmental organizations are welcome to develop and maintain more specialized web-based clearinghouses or databases to meet their specific needs.

a. Case Studies of State Laws, Regulations, and Tariffs

The National Action Plan recommends the development of case studies that provide detailed information about demand response laws, regulations, and tariffs. These case studies could include information on program design, communications/marketing strategies, customer response, and lessons learned. In addition, detailed information on existing state laws, regulations, and tariffs related to demand response programs and dynamic pricing could be added.

Additional contextual information about a demand response program also would assist stakeholders in adapting state laws, regulations, and tariff language to their circumstances. A coalition, or any private or non-federal governmental organizations, could prepare these case studies and DOE will facilitate the inclusion of the material in the Smart Grid Information Clearinghouse.

b. Database of Pilots, Programs and Markets

The National Action Plan emphasizes the importance of developing a database of both pilot and fully implemented demand response programs to include in a web-based clearinghouse, or to be a stand-alone web-based database. When available, the database should include case studies about lessons learned, as well as the impacts, benefits, and drawbacks of the pilots or programs.

A coalition, or any private or non-federal governmental organizations, could be responsible for collecting the information and for developing any analysis included in the database. DOE will facilitate the inclusion of the material in the Smart Grid Information Clearinghouse, while noting in the Clearinghouse that inclusion of non-Federal material does not convey any endorsement or vetting by DOE of the voluntarily-supplied material.

c. Quantitative and Qualitative Summaries

The National Action Plan stresses the value of making detailed quantitative and qualitative information about demand response programs available through the Action Plan's web-based demand response clearinghouse. DOE will be collecting a significant amount of qualitative and quantitative information as part of the demand response projects funded through the Smart Grid Investment Grants, and DOE currently plans to make much of this information publicly available.[15] DOE could assure quality control over any summaries that a coalition, or other private or non-federal governmental organizations, could prepare, and make these summaries widely available through the Smart Grid Information Clearinghouse. In addition, FERC staff will facilitate the inclusion of such information in its own surveys on demand response penetration into the Smart Grid Information Clearinghouse.

d. Measuring Program Results with Criteria and Metrics

The National Action Plan recognizes that there are few criteria or metrics that stakeholders can use to evaluate or compare demand response programs, and thus calls for the development of a standard set of metrics. DOE has developed guidance on how to estimate the benefits of its Recovery Act-funded Smart Grid Investment Grants. In some cases, this guidance could be adapted to support efforts to evaluate the impact of appropriate demand response programs as well. In addition, the National American Energy Standards Board (NAESB), working with stakeholders, has begun to develop widely accepted demand response measurement and verification methods that will help accomplish this activity. A coalition, or other private or nonfederal governmental organizations, could coordinate to develop a broader set of metrics that DOE can include in the Information Clearinghouse.

e. Information for Program Implementation

The National Action Plan highlights the need for up-to-date information about designing and implementing demand response. To date, the impact of demand response programs has not been well understood. While industry trade associations and consultants have provided rate design and program catalogue information, the lack of long-term demand response programs has hindered efforts to analyze their long-term impact. A coalition, or other private or non-federal governmental organizations, could support and facilitate the efforts to provide results from ongoing pilots and programs conducted by industry trade associations and consultants, as more information is available.

f. Online Message Board Capability

The National Action Plan recommends creation of an online message board for qualifying state agencies, regulatory commissions, state energy offices and consumer advocacy groups to exchange information and request assistance on demand response issues, and communicate directly with experts on these matters. This online message board can be included as part of the Information Clearinghouse.

g. Information About the Development of Standards and Protocols

The National Action Plan recommends creation of a centralized online location that stakeholders can visit for information about the development of standards and protocols relevant to demand response and the smart grid. These standards and protocols are crucial to efficiently expand the deployment of demand response; however, different organizations are responsible for the development of the various standards and protocols, making it difficult for stakeholders to follow their progress. A central location that stakeholders could visit would make it easier for them to follow the different processes and, if appropriate, to participate in those processes as well.

Information regarding these standards and protocols is presently housed on the National Institute of Standards and Technology (NIST) website, and the web-based clearinghouse will link to NIST's website and other standards development organizations' websites. FERC staff will commit to work through NIST and the relevant standards development organizations (such as NAESB) to maintain an up-to-date roster of adopted standards with links to on-going standards development processes. A coalition, or other private or non-federal governmental organizations, could assist in the provision of information about the development of standards and protocols.

2. Develop or Enhance Demand Response Estimation Tools and Methods

The National Action Plan calls for the development or enhancement of analytical tools and methods that will: 1) help in the expansion of existing demand response programs and in the creation of new programs by evaluating each program's benefits and costs; 2) advance the use of demand response to support reliable and efficient operations of wholesale transmission, energy, capacity, and ancillary services markets; and 3) better enable end-use consumers to optimize savings, possibly through set-and-forget types of automation not requiring human intervention, while enrolled in a demand response program.

a. System Impact Estimation and Prediction Tools

The National Action Plan highlights the need for the development of tools and methods for conducting system impact estimation and prediction. The information gained from these tools and methods will be used to facilitate a deeper understanding of how demand-side programs can be accurately integrated into modeling frameworks. System impact estimation and prediction tools will allow stakeholders and policymakers to estimate the effects of demand response programs, products, technologies, and incentives on system resource requirements, wholesale prices, and emissions of carbon dioxide and other air pollutants before spending money on pilots and/or full implementation.

FERC staff has developed and made publicly available a tool for estimating the state-level potential for demand response, as well as a quantitative estimation tool that yields information on plant dispatch and energy margins, reserve margins, and unused supply.[16]

All of the Smart Grid Investment Grants, including those awarded for demand response-related projects, include Recovery Act reporting requirements for the grant recipient to collect detailed data. DOE has worked with the smart grid grant recipients to ensure each has a detailed data analysis plan for their SGIG project data, which will be collected and reported to DOE between 2011 and 2013 and made publicly available. Using all the reported data, DOE will then conduct a meta-analysis of all of the individual Recovery Act project results to examine how smart grid technology impacts grid performance, whether positive or negative. It is quite likely that such a large collection and analysis of smart grid project data will be very useful for the development of tools and methods for conducting system impact estimation and prediction.

In addition, demand-side assumptions and impacts are being examined as part of the transmission planning efforts that DOE is funding with Recovery Act funds in each of the three electric interconnections. For example, in the Western Interconnection, the Western Governors' Association's State-Provincial Steering Committee is examining the methodology behind the creation of load forecasts used to support the Western Electricity Coordinating Council's Transmission Expansion Planning Policy Committee production cost modeling. A particular focus is on how these load forecasts account for state and provincial energy efficiency and demand-side management programs and goals. In addition, the State-Provincial Steering Committee is developing a High Demand-Side Management Scenario for possible use in the West's interconnection-wide transmission analyses and planning. As a side benefit, any lessons learned or methods/tools developed in these processes can inform future demand response program as well as resource and transmission planning performed in other fora.

Beyond these federally-sponsored efforts, a coalition, or other private or non-federal governmental organizations, could develop tools and methods for conducting system impact estimation and prediction as well as promote the sharing and posting of tools developed by its members or others.

b. Cost-Effectiveness Tools

The National Action Plan recognizes the importance of cost-effectiveness tools that help stakeholders determine the most efficient mix of demand response programs for a given service territory or region. Demand response program cost-effectiveness can be difficult to evaluate because most current demand response proposals are deeply integrated with broader smart grid initiatives such as advanced metering, meter data management, and complex rate designs. Thus, it is difficult to identify which costs and benefits can and should be specifically attributable to the demand response program, including the separation of those costs and benefits from the rest of the smart grid initiative's costs. Another challenge is the long-standing issue of projecting longterm impact and benefit streams (such as avoided resource costs) from a demand response program.

DOE has funded the development of a cost-effectiveness screening tool for the Pacific Northwest Demand Response Project (PNDRP) that other organizations and regions could use. DOE also has established guidance on the identification of smart grid program metrics and impacts.[17] Instructive materials developed by the Electric Power Research Institute (EPRI) and several private consulting firms are available for proprietary use. However, since cost-effectiveness testing requires a better understanding of the impacts of demand response programs, more data and analysis are required (as discussed above) before better cost-effectiveness tools can be run using well-founded impact and benefit estimates.

Cost-effectiveness tools and rules must meet the needs and requirements of state regulators, reflect the structure of retail and wholesale markets for a region, and meet the needs of utilities that request demand response program recovery in ratemaking proceedings. DOE commits to work with the state regulatory community to determine whether this community is willing to consider and use federally-sponsored research into demand response cost-effectiveness guidelines and tools.

In addition, a coalition, or other private or non-federal governmental organizations, could consider the development of new methods or tailor existing methods for analyzing the effectiveness of different demand response programs and rates.

c. Decision Tools and Educational Materials

The National Action Plan calls for the development of decision tools for end-use customers, including customers with special circumstances, such as customers that manage multiple facilities. In addition, there is a definite need for educational materials to support customer understanding.

The National Action Plan identifies these tools as crucial to help end-use customers understand how to change their consumption patterns to achieve bill savings and provide system benefits while enrolled in a demand response program.

Customer evaluations about the value of demand response offerings depend on several factors, such as the type of demand response program or rate, the cost of participation (hardware, software, and internal operations and staffing costs or opportunity costs), and the potential benefits or payments accruing from participation.

Furthermore, demand response options are becoming more diverse every year, with a variety of technology and operational requirements, offerors, curtailment or management strategies, and financial and reputational benefits.

While large, multi-site companies may have developed their own tools and materials to assist in evaluating and participating in demand response programs, purchasing products, and employing technologies, smaller customers may need help understanding their electricity usage profile, available options to control their use, and means to assess cost-effectiveness.

A coalition, or other private or non-federal governmental organizations, could identify new tools and plans that each customer class and industry needs, as well as support private efforts to develop the tools that are identified. In identification of new tools, a coalition, or any private or non-federal governmental organizations, can leverage tools developed by DOE. One example is the benchmarking tools for buildings sponsored by DOE and EPA as part of the ENERGYSTAR® program.

d. Cost Recovery Tools and Methods for Enabling Technology

Many types of demand response programs require the adoption of enabling technologies such as advanced metering, meter data management, and—to maximize demand response impact—customer in-premise automation and analysis tools.

The National Action Plan assumes that utilities offering demand response programs that require enabling technologies will seek to recover the costs of the enabling technologies in their rates. The National Action Plan therefore emphasizes the need for tools and methods that utilities and state and local governing officials can use to develop a business case and examine options for recovering costs of installing approved, cost-effective new technologies that enable demand response. Such tools will help utilities and state and local governing officials estimate the financial implications of demand response options and investments.

For the past five years, utilities have been designing and submitting business cases for a suite of smart grid investments, with varying degrees of success. To assist utilities in developing business cases, an industry of consultants has emerged and a library of reports and regulatory filings are now available.

At present, most of the utilities winning Smart Grid Investment Grants have submitted their business cases and cost recovery proposals to regulators. DOE has developed guidance for grant recipient reporting on metrics and benefits.

e. Tools and Methods for Incorporating Demand Response in Dispatch, Ancillary Services, and Transmission and Resource Planning

The National Action Plan calls for the development of new tools and methods that can directly incorporate demand response into dispatch algorithms and resource planning models.

The National Action Plan also highlights the need for better methods to forecast and model the capability of demand resources to adjust consumption in near real-time and for hardware and software tools that allow customer appliances or vehicles to autonomously provide reliability services, such as frequency response service. These tools and methods will allow users to better capture the capability of demand response to serve as an alternative to building new generation and transmission and to act as a resource to alleviate transmission congestion.

In parallel with assessing the amount of demand response available, the nation's grid operators have been studying alternative ways to incorporate demand response into system dispatch and planning.

However, their ability to forecast and incorporate demand response for near-term operations varies as a function of the type of demand response. Specifically, reliability- or event-driven, dispatchable or externally-controlled programs are more easily forecasted than demand response offerings that are price- or rate-driven and subject to customer decision and action rather than fully automated.

Experience gained with the Smart Grid Investment Grant-associated demand response and variable rate projects should offer insight into how customers react to different types of demand response offerings and how different offerings affect energy use levels. Thus, in several years there should be enough data to create accurate algorithms and methods for forecasting demand response impacts on loads, and for incorporating different types of demand response resources into grid operations. DOE intends to work with FERC staff, grid operators, demand response providers, and other interested organizations to help create such tools when sufficient underlying data become available.

In addition, the entities being funded by DOE through the Recovery Act to conduct interconnection-wide transmission planning will need to develop tools and methods to incorporate demand-side measures into their plans, to meet their own requirement to better include the demand side. Such efforts are currently underway.

A coalition, or any private or non-federal governmental organizations, also should assess the need for a coordinated and joint project to develop a means of incorporating demand resources in wholesale electricity markets and in resource planning processes.

f. Transparency of Price Information

The National Action Plan recognizes that for the United States to realize its full demand response potential, electricity customers must have access to, and a better understanding of, information about real-time or near-real-time energy prices.

Depending on the type of customer and demand response program or provider, the customer also may need information about ancillary service prices, capacity payments, and congestion costs. Better price information delivered more clearly will help potential demand response providers design market offerings, assist utilities in designing demand response-encouraging rates, and help potential demand response customers evaluate whether to participate in a demand response program.

While the National Action Plan recommends development of tools that provide access to transparent price information, FERC staff and DOE believe that an important first step toward developing better tools is to first provide better price information. NIST and NAESB are currently examining and developing standards for the provision of energy price information.

In addition, roughly two-thirds of the Smart Grid Investment Grants that include some form of demand response will explore a variety of means to communicate information about the value of electricity to customers, and when and how customers should change their energy use behavior. DOE's ARPA-e program also has awarded approximately $5 million to the Precourt Institute for Energy Efficiency at Stanford University for the creation of a system that combines behavioral techniques with human-centered design and technology to achieve large-scale energy reductions. At the same time, commercial demand response providers and the smart grid technology community are developing a wide variety of new tools and technologies to help customers respond. A coalition, or any private or non-federal governmental organizations, could oversee private efforts to assess the need for a coordinated and joint project on transparency. If a joint project is considered beneficial, a coalition, or any private or non-federal governmental organizations, could coordinate the development of guidelines on transparency.

APPENDIX A: RESEARCH TOPICS

Appendix A of the National Action Plan included a list of areas that may benefit from further research. While FERC staff and DOE believe that it is appropriate for the federal government to conduct research on some of those topics, the list below includes other research studies that, in some cases, may be more appropriately performed by individual technology vendors, demand response providers, and utilities that will directly benefit from the research.

Social Science Research Issues

- Exploration of issues related to rate design, metering costs, and an analysis of customer response to time-varying prices in successful programs.
- Understanding the interest of elected officials, regulatory agencies, utilities and load-serving entities, RTOs/ISOs, other demand response providers, and third-party vendors in making demand response a priority, or not.
- Sponsoring the development of a comprehensive database of information on demand response programs. This database would be designed to allow the analysis of the following issues:
    - How the participation rate and demand impacts vary depending on whether the demand response rate design is required by law or regulation (e.g., as the default rate) or if customers voluntarily sign up for it.
    - Investigation of typical demand response participation rates and how these rates vary based on program design parameters, marketing strategy, customer type (residential, commercial, and industrial; gender and income level), and customer profile (e.g., age, geographic location). Included in this investigation should be an examination of the non-price response incentives that motivate customers to participate in demand response programs and products.

- Exploration of the effectiveness of demand response when critical days are sustained for multiple days in a row, and the persistence of demand response behavior when programs and products are implemented over a long-term time horizon.
- Understanding how consumers shift demand to off-peak hours of the day and if consumers take conservation measures to lower peak demand over all hours of the day, not just during peak hours.
- Impacts of customers participating in multiple demand response programs.

Technical Research Issues

- Relative benefits and costs of various types of demand response programs, products, technologies, and incentives, covering such topics as the various types of benefits, potential costs, and paybacAn analysis of the positive and negative environmental effects of demand response, including but not limited to an estimate of emissions mitigation when considering load reduction versus load shifting.
- An assessment of the potential for demand response to assist in the integration of variable resources such as some forms of renewable energy. Included in this assessment would be: a study of how demand response resources can be dispatched to support and balance variable generation from some forms of renewable energy; and a calculation of environmental benefits stemming from the reduced need to build new generation capacity and transmission systems.
- A study of how long demand response resources can be expected to provide resource adequacy and reliability benefits compared with other resources, such as generation, transmission, and storage.
- A study of how plug-in hybrid electric vehicles (PHEVs) interact with demand response programs, examining whether demand response rate design provides a price signal that encourages PHEVs to charge during off-peak hours, as well as how different demand response pricing mechanisms interact with PHEVs and their net impacts on how demand for electricity might change.
- Research into the appropriate codes and standards (e.g., building codes, consistent definitions, performance standards) for demand response, considering whether they are mandated at the federal or state level; whether they are mandatory or voluntary requirements; how they will be complied with and enforced; and how the costs

involved in their development, compliance, and enforcement will be paid for. Integration of the many utility and third-party programs, devices, and software applications needs to be taken into consideration.

- Implementation of demand response programs and products at campuses owned by municipal, county, state, and the federal government, covering topics such as the types of demand response suitable for government customer applications, barriers and lessons learned from existing programs, percentages of the state electricity portfolio under municipal, county, state, and federal control, and the potential impact of incorporating government loads into demand response programs and products.
- Best strategies to coordinate state and local-regulated retail demand response programs with RTO/ISO demand response programs for organized power markets.
- Curtailable load programs aimed at industrial customers, covering topics such as potential benefits to all customers, cost recovery, rate design, and program marketing.
- The impact of distributed generation as a resource that can be integrated more effectively in conjunction with demand response.k horizons associated with each.

APPENDIX B: LETTER FROM NATIONAL ACTION PLAN COALITION

NATIONAL
ACTION PLAN
COALITION

The Honorable Steven Chu
Secretary, U.S. Department of Energy
1000 Independence Ave., SW
Washington, DC 20585

The Honorable John Wellinghoff
Chairman, Federal Energy Regulatory Commission
888 First Street, NE
Washington, DC 20426

Dear Mr. Secretary and Mr. Chairman:

I am writing on behalf of a number of non-profit organizations who are stakeholders in the area of demand response and smart grid. I am writing relative to the National Action Plan on Demand Response (NAP), which the Department and the Commission are charged by Congress with implementing.

When your agencies issued the NAP in June of 2010, one of the Plan's recommendations was that a "Coalition of Coalitions" be formed to implement it. That Coalition has indeed been formed and is ready to work with you to begin this implementation.

The NAP Coalition was formed immediately upon the release of the Plan and has been meeting biweekly since that time. The Coalition has developed a work plan and already begun several work efforts, including developing case studies, conducting webinars and developing content and activities related to the communications/education component of the Plan.

The NAP Coalition continues to add members. At present, the organizations working together are:

- Alliance to Save Energy (ASE)
- American Council for an Energy-Efficient Economy (ACEEE)
- American Public Power Association (APPA)
- Demand Response Coordinating Committee (DRCC)
- Demand Response and Smart Grid Coalition (DRSG)
- Edison Electric Institute (EEI)
- Environmental Defense Fund (EDF)
- GridWise Alliance
- National Association of Regulatory Utility Commissioners (NARUC)
- National Association of State Energy Officials (NASEO)
- National Association of State Utility Consumer Advocates (NASUCA)
- National Rural Electric Cooperative Association (NRECA)

The NAP Coalition stands ready to expand and accelerate its efforts related to the implementation of the Plan. One member of the Coalition, DRCC, has contributed initial "seed" funds to support the current work effort. Beyond that, Coalition members are prepared to combine their expertise and financial resources - in-kind and monetary - with those of DOE and FERC. Members are prepared to enter into an agreement for a public-private partnership that will rapidly accomplish the goals of the National Action Plan and support achievement of the Administration's goal to modernize the nation's electricity system and capture the benefits that can be realized from such.

1615 M St NW, Suite 900, Washington, DC 20036

I would be pleased to provide additional information or answer any questions you may have about the National Action Plan Coalition. Thank you for your support and consideration.

Best regards,

Dan Delurey

Dan Delurey
National Action Plan Coalition

1615 M Street NW
Suite 900
Washington, DC 20036
202.296.1686
dan.delurey@demandresponsecommittee.org

End Notes

[1] Energy Independence and Security Act of 2007, Pub. L. No. 110-140, § 529, 121 Stat. 1492, 1664 (2007) (to be codified at National Conservation Policy Act, 42 U.S.C. §§ 8241, 8279).

[2] Id.

[3] As discussed infra, FERC and DOE are not endorsing in this Implementation Proposal a particular coalition, and note that entities may form, or have been formed, to pursue activities that may involve aspects relating to the National Action Plan.

[4] A National Assessment of Demand Response Potential (June 2009), available at http://www.ferc.gov/legal/staffreports/06-09-demand-response.pdf (National Assessment).

[5] Id.

[6] National Action Plan on Demand Response (June 2010), available at http://www.ferc.gov/legal/staff-reports/06- 17-10-demand-response.pdf (National Action Plan).

[7] EISA section 529(a).

[8] National Action Plan, supra note 5.

[9] EISA section 529(c).

[10] Note that the headers used in this Implementation Proposal are primarily taken from and are consistent with the National Action Plan.

[11] As an example of the type of agreement that EISA section 529(c) directed for inclusion, Appendix B contains a letter from several entities stating that they have formed a Coalition to implement the National Action Plan. This implementation proposal does not endorse a particular coalition, and notes that other entities also may form, or have been formed, to pursue activities that may involve aspects relating to the National Action Plan.

[12] The Lawrence Berkeley National Laboratory is supporting the two interconnection-wide entities in the west (Western Electricity Coordinating Council and the Western Governors' Association), while Oak Ridge National Laboratory is supporting the two interconnection-wide entities in the east (Eastern Interconnection Planning Collaborative and the Eastern Interconnection States' Planning Council). The fifth entity, the Electric Reliability Council of Texas, was also awarded Recovery Act funds for interconnection-wide planning and analysis and will also be considering demand-side scenarios.

[13] DOE's technical assistance is defined to be largely on policies and programs for states, regions, independent system operators, and sometimes utilities that wish to begin or improve on their demand response activities. DOE's technical assistance on demand response increasingly is integrated with its increasing technical assistance to states on smart grid, particularly as the subjects of demand response and smart grid are closely related. It is noted that DOE's technical assistance on demand response (and smart grid) is part of a larger commitment by the DOE Office of Electricity Delivery and Energy Reliability to provide technical assistance on any electricity policy requested by state and regional officials. As such, DOE's assistance for demand response issues may at times compete for available resources devoted to technical assistance to states with non-demand response electricity policy subjects.

[14] The Smart Grid Information Clearinghouse includes information about demonstration projects, research and development, best practices, standards, legislation, and regulation. DOE awarded the Virginia Polytechnic Institute and State University a five-year $1.25 million grant to develop and maintain the Clearinghouse. See Barbara L. Micale, $1.25 Million Grant To Take Lead on Smart Grid Information Clearinghouse Web Portal Awarded to Virginia Tech, Virginia Tech News, October 27, 2009, available at http://www.vtnews. vt.edu/story.php? relyear=2009& itemno=801.

[15] Information regarding customer electricity usage will be masked to ensure customer privacy.

[16] The tool for estimating state-level potential is available at http://www.ferc.gov/industries/electric/indusact/demand-response/dr-potential.asp.

[17] Guidance is located at http://www.smartgrid.gov/files/teams/metrics guidebook.pdf.

INDEX

D

E

F

G

Q

R

S

T

U

V

W

Y